INDEPENDENT ~~~~~~~~ ES
OF T[

ALAN NALDRETT

Kissimmee Automobile Company, c. 1907

Independent Car Companies of the USA

Alan Naldrett

Independent Car Companies of the USA / Alan Naldrett
1st Edition Paperback: May 2020
ISBN: 978-1-940676-47-0
Cover Design: Chance Cooray
Editor: Shoshana Horowitz
Interior Design: Susan Veach

Published by Charles River Press, LLC
Printed in the United States of America

TABLE OF CONTENTS

This book is dedicated to

Bob & Evelyn Cole, Greg and Sue Cole, Kent and Michelle Cole,

Ron and Norma Naldrett, Frank and Sherrel Hissong, Kirsten

Gardner, Tom and Lindsay Hartley, Michele Pratt, Gail Zabowski,

Lynn Lyon, Dave Keck, Dave Castle, Ellen and Bill Carpenter,

Nathan and Regina Hissong, Dean Naldrett,

Jenna and Mike Karlick, Hannah Naldrett, Jammie Naldrett,

Rick and Rosa Margarita Rasmussen, Craig, Brian, and

Tom Nienhaus, Basement John Shinnick, Terry Ernest,

Carl Moss, and all my awesome Cole cousins!

Acknowledgements-

I wish to acknowledge C. Harold Wills, Beverly Kimes, Robert Hupp, Ransom Olds, James Packard, Roy Chapin, Howard Coffin, Henry Joy, Fred Duesenberg, and all the other unsung car heroes without whom we wouldn't have the cars and auto technology we have today.

INTRODUCTION

IWANTED TO WRITE ABOUT THE many independent car companies in the United States and knew that quite a few existed, but I soon realized that the number of companies exceeded my imagination. One of my goals was to find an auto company for each state. Two states gave me trouble; neither Nevada nor Utah has ever had an automobile company located within their state boundaries (so far), but those two states did have tales of scam artists taking advantage of the automobile furor that occurred between the late 1890s and 1919: back in the early twentieth century, so many people wanted to be involved in anything that had to do with autos and the auto industry that it was easy for a smooth-talking con artist to come to town, feed on the residents' dreams, sell stock, and then disappear.

Thirteen states had only one car company created within its borders. These "one-car wonder" states were Alaska, Arizona, Arkansas, Delaware, Georgia, Hawaii, Louisiana, Montana, New Mexico, North Carolina, Oregon, Vermont, and Washington.

Of the remaining states, the number of car companies they each had is almost incalculable. Indiana had at least fifty-five *notable* car companies, Ohio over sixty companies, New York over thirty, Wisconsin over twenty, and Michigan over two hundred. Other states with multiple auto companies include Massachusetts, Illinois, Pennsylvania, and Minnesota.

A lot of companies are well known and have enough information available about them to do a whole book on each company. Dozens of books have been written about the Big Three companies: Ford, Chrysler (and whichever company they're partnered with), and General Motors. Many books are also available on the

Little Four companies—Nash, Hudson, Studebaker, and Packard—as well as many other auto companies, including Knox, Hanson, Willys, and Duesenberg. Because of those books and other information available about the major car companies, I have not gone into as much detail about them so as to make room in this book for information and stories about the more obscure, but viable, car companies.

A lot of companies were "regional companies" that only sold their vehicles within a certain territory (like Arkansas's Climber and California's Kleiber). Every state had at least one hardy individual who liked to tinker with building autos, but not all of them started companies to mass produce their vehicle, although that was often their dream. Many of these companies weren't from a large city, although many were. The car maker was just as likely to be a rural dweller, out on the farm, as to be a big city slicker. However, a rural person would usually need to move to an urban area for labor and access to materials in order to keep their company and dream alive beyond a few years.

What happened to all those car companies? The most prevalent reason that they went out of business was that most of the companies weren't capitalized enough and therefore didn't have the money to produce enough cars to make a profit. Usually, the companies that were successful were capitalized at a minimum of half a million dollars.

Many companies sold out to larger companies. Scripps-Booth Company had been a low-producing car company but made a lot of money for its founder when General Motors purchased it. Of the companies that produced cars, many declared bankruptcy, were forced into receivership to pay creditors, merged with other companies, were purchased by other companies, and sometimes just ceased operating, disappearing into the dead of the night. Many of the companies continued to manufacture just one or two car components such as transmissions, batteries, or door handles. As many stories as car companies exist; these are just a fraction of them.

Auto assembly line

1920s auto factory

FOREWORD

THE UNITED STATES HAS HAD over a *thousand* (some sources say two thousand) auto companies that produced cars at one time or another. Why so many? Because autos were one of those fanciful inventions that many young men, especially at the dawn of the twentieth century, would dream about at night. In the daytime, many of them would obsessively draw cars and engine diagrams. As Americans heard about the advances in auto manufacturing being made in Europe, this made the dreams seem more possible. Manufacturing a car was possible, and the ultimate dream was to own a *car company!*

From the 1850s through the 1890s came reports of various inventors and innovators in the United States who had successfully tested motorized vehicles. Could it be financially possible to *manufacture* motor cars and show a profit? Slowly but predictably, cars began to be produced and sold in the United States, starting on the East Coast.

From the 1890s to the 1910s, the New England states dominated the automobile field with dozens of companies, mostly ones that manufactured steam and electric motor vehicles. This included Connecticut, which had the number-one-selling (from 1897 to 1900) motor vehicle in the United States: the Columbia electric auto. Massachusetts was also a New England state with several early car companies.

By 1903, as Detroit began its ascendancy to the top of the auto world, the East Coast states of Massachusetts and Connecticut made their descent. After the East Coast was no longer on top, two areas became the top auto-producing areas: the state of Indiana (particularly Indianapolis) and Cleveland, Ohio. After the heyday in the 1900s of Indianapolis and Cleveland, in 1910 and beyond, Detroit became the place with the most auto manufacturers.

As Detroit became the Motor City, surrounding cities in Michigan caused it to become the dominant auto state with motor vehicles built in Kalamazoo, Flint, Pontiac, Grand Rapids, Port Huron, Marysville, Alpena, Fenton, and others as well as Detroit.

Other states with an early major auto company presence included Wisconsin with *Jeffery, Rambler, Excalibur, Mitchell,* and **Kissel Motor Companies** and New York with *Pierce-Arrow* and *Franklin.* Chicago, Illinois, and the rest of the state of Illinois had several companies as well as Pennsylvania. Many attempts were made to start auto companies all over the nation. A greater number of "paper companies" never had any auto production than the number of companies that actually manufactured cars or had a prototype. However, MANY car companies did manufacture a lot of different autos over the years. Many variations and types of vehicles were made in the fifty states and Washington, D.C., and many of their stories are recounted here.

Throughout the text, the car companies are printed in **bold** type. The car names are in *italics.* Photos are from the author's collection, shot by the author, or in the public domain unless otherwise noted.

CHAPTER 0:

THE BIG THREE: GENERAL MOTORS, FORD, AND CHRYSLER

IN 1950, THE LITTLE FOUR car companies—**Hudson, Packard, Nash,** and **Studebaker**—disappeared as major competition for the Big Three: **General Motors, Ford,** and **Chrysler**. A lot of the damage can be blamed on the price war on passenger autos between Ford and Chevrolet (GM) that brought the prices down so low it became hard for the independent companies to make a profit.

Because of the price wars and the rising cost of doing business, **Hudson** and **Nash** merged to become the **American Motor Company**, which was later absorbed by **Chrysler**. **Packard** became part of **Studebaker**, and both completely disappeared by 1963. Thereafter, only the Big Three remained in America until the late 1970s and the rise of foreign auto companies, including **Toyota** and **Honda**. The first non-American factory in the United States was built by **Volkswagen** in 1978.

It took three tries for Henry Ford to form a successful auto manufacturing company, but the third time was definitely the charm. The **Ford Motor Company** was the top-selling car company even before their iconic Model T, the *Tin Lizzie*, came out. **Ford** spent over thirty years in the number-one position!

Before he started **General Motors**, William "Billy" Durant, with his partner J. Dallas Dort, was one of the most successful carriage builders in the United States. Billy Durant bought the companies of various carriage parts—such as carriage wheels—and moved them to Flint. He took much the same approach when he decided to purchase the **Buick Motor Company.** He bought other auto companies and auto-parts companies and consolidated them into **General Motors.**

The **Maxwell Motor Company** was a successful firm for over fifteen years, but when they got into financial trouble, they got Walter Chrysler to help them out. He did, buying out the company and naming it the **Chrysler Motor Company** after himself.

The Maxwell factory in Indiana

American Motors Company (AMC) headquarters in Detroit

Chapter 1:

The First (and Only?) Car Built in Alaska: The Sheldon

NFORTUNATELY, ALASKA IS PROBABLY THE worst example of a state to start off with when detailing the stories of each state's car companies because only one auto has been manufactured there: the *Sheldon*. This was the creation of Robert Sheldon, who had moved to Skagway, Alaska, with his family during the Klondike Gold Rush of 1899. In 1905, he put together a gasoline buggy that became the only car ever built in Alaska (so far). He would use this car, now on exhibit at the museum of the University of Alaska in Fairbanks, to tool around the mining areas; this helps to explain why he was a road advocate, campaigning for better roads. He served Alaska as postmaster and in the Alaska State Legislature before dying at the age of ninety-nine.

Robert Sheldon in his car, the Sheldon

Chapter 2:

The Premier Car of Alabama: The Premocar

ONE OF THE FEW CARS manufactured in Alabama was the *Premocar* of the **Preston Motor Corporation** from approximately 1920 to 1925, which was named after early financier Preston Orr. In 1918, W. H. Tarpley designed the car, named the *Premocar,* and the company opened with a factory at 18th Avenue and Vanderbilt Road in Birmingham, Alabama.

The *Premocar* was offered in two types: a four-cylinder model and a six-cylinder model. The six-cylinder model had a Falls OHV engine and was priced lower than the four-cylinder model, which had a Rochester-Duesenberg engine. The four-cylinder model was dropped after the first year, and most of the remaining *Premocar*s built had six-cylinder engines.

President Harding in a 1921 Premocar

In 1921, a white *Premocar* touring car was used to drive President Warren Harding around on a tour of Birmingham, Alabama. A later deluxe version of the *Premocar* was also used to drive President William Howard Taft and John Philip Sousa around the city.

*Factory and employees of the **Preston Motor Corporation***

The company's fortunes were at a low point when three of the board members were indicted for violation of Alabama's Blue-Sky Stock Laws in 1923. The company eventually went into involuntary receivership, and the factory was auctioned off. It was estimated that a little over three hundred *Premocars* were built from 1920 to 1925.

Chapter 3:

The Climber of Arkansas

THE **CLIMBER MOTOR CORPORATION** WAS from Little Rock, Arkansas, and manufactured the auto the *Climber* from 1919 to 1924. It was an open car with a Herschel-Spillman engine and was available in four- or six-cylinder versions.

In 1919, the firm purchased 20½ acres on the east side of Little Rock and built their auto plant. By the end of the year, they had ten distribution agencies set up, mostly in Arkansas. The car was offered in touring and roadster styles. The *Climber* executives claimed to be working on commission only. Chief engineer George Schoeneck, who had been imported from Detroit, said, "We are not making promises; we are making cars," but dissension hit the ranks the next year as parts deliveries ran slow, the three thousand cars promised for 1920 was no longer possible, and a few of the partners quit.

For 1923 and 1924, only the six-cylinder model was offered, and a coupe style was added, but the company ran out of funds and was sold to a group who continued to assemble *Climbers* from the remaining parts as the **New Climber Company.** When the parts were gone, the company was dissolved. The stockholders received back 88.4 percent of their investments.

1923 Climber from Little Rock, Arkansas
(photo courtesy of Wikipedia)

Chapter 4:

Arizona's Copeland

Louis D. Copeland made his first attempt to motorize a vehicle while he was living in Arizona in 1881. He unsuccessfully tried to put an engine on a Columbia bicycle. Further attempts were better, and he created a "steam bicycle" that was propelled by a little engine attached to it. Although Copeland found a backer in **Northrup Industries** of Newark, New Jersey, he failed to sell enough to make a go of it. Retiring to California in the 1890s, he concluded that no one would *ever* be able to make a profit selling the public an automated vehicle.

Chapter 5:

California's Multitude of Car Companies

ALIFORNIA HAD MANY CAR COMPANIES, and lots of them had "California-style" brand names. One of the first motor cars in the state was from **A. Schilling & Sons** in San Francisco. They produced the *Golden Gate* automobile from 1894 to 1895. It was a three-wheeled, two-seater car with a two-horsepower engine. One was sold to a lady in Santa Maria in 1894 and was the first sale of a gasoline vehicle in California.

The **Kleiber Motor Car Company** in San Francisco was known for the commercial vehicles they built. From 1924 to 1929, they produced a passenger car with a six-cylinder engine that was only marketed on the West Coast. Paul Kleiber was the founder of the car that was advertised as "aristocratic in appearance."

1925 Kleiber Four-Passenger Coupe

The *Kleiber* auto was plagued by low production, producing less than one thousand autos in the five years they were manufactured. Although they quit making autos in 1929, they continued making commercial vehicles until 1937.

———————

The **Durocar Manufacturing Company** built the *Durocar* in Los Angeles from 1906 to 1911. This two-cylinder, twenty-six-horsepower, water-cooled car had three models offered: surrey, runabout, and touring body. The four-stroke engine was placed under the front seat. A three-speed selective transmission with a drive shaft powered the drive to the rear axle.

The *Durocar* was founded by Walt Moreland, who at first worked as a mechanic for Alexander Winton. Moreland came to L.A. to work on some *Wintons* and never left. He at first worked on the *Tourist* and the *Magnolia*. He met the Long Beach dealer for the *Tourist*, William M. Varney, and started the **Durocar Manufacturing Company** with him in 1906.

The car was introduced in January 1907 at the Los Angeles

Auto Show. Production began in February, but by September, only seven cars had been built. After factory renovations, production rose to five cars a day. Even though the car was built in Los Angeles, the advertising gave a nod to San Francisco with this motto: "The car with the motor for San Francisco's hills."

When the Board of Directors refused to manufacture a four-cylinder car in 1908, Moreland quit. The company DID produce a four-cylinder car in 1910, but the company wanted to reincorporate. Moving to a new factory in Santa Ana, a few autos were produced from parts, but then, the company went under.

1911 Durocar (photo courtesy of the San Diego Historical Society)

One of the longest-lasting and best-known car companies of California was the Los Angeles-based **Auto Vehicle Company.** They sold the *Tourist*, which was offered in two- or four-cylinder models, from 1902 to 1910. The cars were five-passenger touring models with a water-cooled engine. The two-cylinder models had a friction transmission, while the four-cylinder models had a sliding-gear transmission; they sold for $1,700. The cars were mostly sold in California.

The factory was at 943 North Main Street in Los Angeles. In 1905, when they were producing over five hundred cars a year, they moved to larger quarters at Tenth Street and South Main in Los Angeles and started manufacturing trucks as well. In 1908, they assembled a hose cart and chemical wagon for the Hollywood Fire Department. In 1909, the company was purchased by the **California Automobile Company**, who continued making both the *California* and the *Tourist* cars in 1910.

At the end of 1909, the company announced that they were having trouble getting parts and would stop manufacturing vehicles, instead continuing as a car dealership. As a dealership, they sold *Firestone, Columbus, Warren-Detroit,* and *Columbia Electric* vehicles until 1914, when they returned to manufacturing by making the *Beardsley Electric*.

*A Tourist car manufactured from 1902 to 1909 by the **Auto Vehicle Company** (photo courtesy of Metropolitan News Company)*

Changing their name to the **Beardsley Electric Company,** they produced the *Beardsley Electric* from 1914 to 1917. Variety was the key, as *twelve different models* were available! The lightest model did a brisk twenty-eight miles per hour. The cars were equipped with wire wheels and had a ninety-two-inch wheelbase. They sold from $1,285 and up.

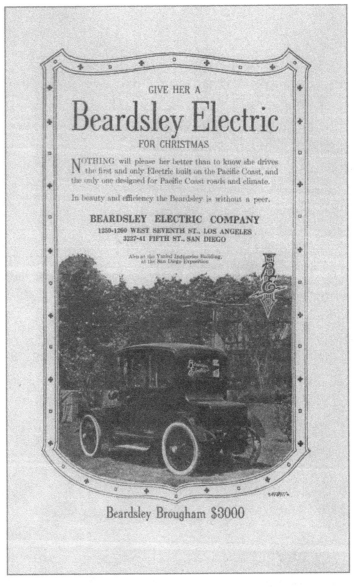

Beardsley Electric ad says, "Give her a Beardsley Electric for Christmas!"
(photo courtesy of Learn California)

CHAPTER 6:

THE COLBURN CAR COMPANY OF COLORADO

THE **COLBURN AUTOMOBILE COMPANY** IN Denver built a variety of models from race cars to large, touring cars with the radiator behind the engine. The touring car was noted for a sloping hood which hid the radiator behind it. The company operated from 1906 to 1911.

The *Colburn* 1909 roadster was advertised to run at speeds of over seventy miles per hour. The cars weighed 2,600 pounds, and the four-cylinder engines were rated at 40 horsepower. The Model 30 and Model 40 often raced to great success in the Denver area. They sold for around $4,500 and were known for a shiny, gold "C" on the front of the autos.

The founders of the company were brothers Herbert and Ernest Colburn Jr., sons of Judge Ernest Colburn Sr. The judge made a fortune with the mines he owned in Cripple Creek, Colorado, and financed the cars until he and his sons had a falling out in 1911 and the factory was closed.

*The Colburn automobile is so rare that none are known to exist today.
The driver is Herbert Colburn; his father, Judge Ernest Colburn,
is to his left.*

1910 Colburn Model 35 Touring Car

CHAPTER 7:

THE TOP-SELLING POPE-COLUMBIA AND LOCOMOBILE OF CONNECTICUT

CONNECTICUT WOULDN'T HAVE BEEN ONE of the pioneer auto states without Albert Pope. Albert Augustus Pope (1843–1909) was the first great American auto entrepreneur. After service in the Civil War, where he was in the Battle of Bull Run, he earned the title of colonel, a title which he used for the rest of his life. Col. Pope's business empire began with the buying and selling of real estate, and then the profits were used to form the **Pope Manufacturing Company.** This company was formed to produce small articles, including air pistols, cigarette rollers, and shoe parts and supplies.

In 1876, Albert Pope attended the Philadelphia Centennial Exhibition and was drawn to the exhibit of English bicycles. Pope imported fifty bicycles and sold them. In 1878, he had his own bicycle constructed and outsourced the manufacturing to the **Weed Sewing Machine Company** in Hartford, Connecticut. The bicycle was the *Columbia*. In 1880, Pope took over the **Weed** bicycle operation. He had been purchasing bicycle patents, and in 1899, he organized the **American Bicycle Company.** This was a consolidation of forty-five bicycle-producing firms. As the bicycle craze started to end, Pope moved on to invest in automobiles.

In 1897, Pope partnered with the **Electric Vehicle Company** to produce the electric car called the *Columbia*, named after Pope's successful bike. The *Columbia* became the top-selling car in the nation for the next three years.

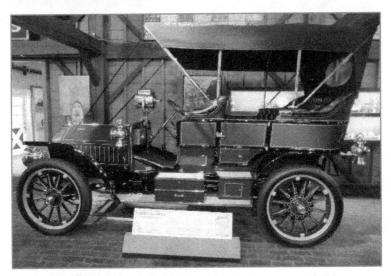

Albert Pope, a Pope-Hartford trading card, and a 1906 Columbia

Pope still produced bicycles along with the automobiles. He knew that each vehicle had a better future if the roads were improved, so in 1900, Pope became a leader in the Good Roads Movement.

The *Pope-Hartford*, produced in the main headquarters in

Connecticut, had a one-cylinder engine, with a two-cylinder engine on the market by 1905 and a four-cylinder engine in 1906. The *Pope-Tribune* was the Maryland division of Col. Albert Pope's **Pope Manufacturing Company.**

In 1903, Pope purchased the **Robinson Motor Vehicle Company**, and it became the **Pope-Robinson Company** of Hyde Park, Massachusetts. The *Pope-Robinsons* were larger than other cars under the **Pope** banner, but their $6,000 price was generally the reason they didn't sell. They only lasted until 1904. The company was purchased by the **Buick Motor Company** in 1904.

Pope had a car for every budget: the *Pope-Toledo* was the more luxurious car, followed by the *Pope-Hartford*, and then the *Pope-Tribune*, which sold for $500 to $900.

A Pope-Columbia trading card.

A 1911 Pope-Hartford

The **American Bicycle Company** was like many other bicycle companies that saw that the bicycle craze of the 1890s was giving way to the Brass Era of autos. They announced in September 1900 that they would henceforth be producing automobiles instead of bicycles. The majority owner Frederick Billings surrendered rights to the **American Bicycle** name and retooled his Toledo factory to produce steam autos. Renamed the **International Motor Car Company,** the company came out with two steamer models called "International" in 1901, with models named the *Toledo* and the *Westchester.* By the end of the year, the entire marque was moved to Toledo.

In 1902, the **International Motor Company** added a three-cylinder, sixteen-horsepower car to its line. Selling fifty of the gasoline models in their first year, in 1903, the company dropped a steam model and added two more gasoline models. The company was sold to Alexander Pope, who continued it as **Pope-Toledo**.

The *Pope-Toledo* car of Toledo, Ohio, was announced in 1904. The **International Motor Car Company** became the **Pope Motor Car Company** with the *Pope-Toledo* as its 1904 offering. Rounding out the cars were the *Pope-Tribune* (1904–1907) of Hagerstown, Maryland, and the *Pope-Waverly* of Indiana. The *Pope-Waverly* was an electric vehicle, renamed when Pope purchased the **International Motor Car Company** of Indianapolis, Indiana. The company made the two-seater electric car the *Waverly* from 1901 and continued with this model until 1914.

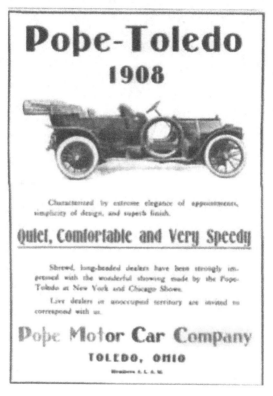

On the West Coast, the Pope car won some significant racing events, which started the marque off in grand style. It was advertised as the "mile-a-minute car" and as the "car that meets every requirement." It had popular features such as copper water jackets and a three-speed sliding-gear transmission. In 1911, a six-cylinder model was introduced. Production was usually about seven hundred vehicles per year.

In 1908, problems began in the **Pope** empire. Albert Pope died on August 10, 1909, and his brother George took over the company. In 1909, the main plant (but not the whole company) was purchased by the **Overland Motor Company** as the **Pope Company** declared bankruptcy. The company was reorganized and lasted until 1915, when it was liquidated.

In Bridgeport, Connecticut, the *Locomobile* was founded in 1899 by John B. Walker, the editor of *Cosmopolitan* magazine. He had purchased half the rights to the Stanley twins' steam car. His other partner, A. Lorenzo Barber, started his own company, the **Mobile Motor Company.**

A steam-driven, early Locomobile and a later, gasoline-powered Locomobile

1914 gasoline-powered Locomobile Model 48

Both partners discovered that the autos with the steam-car design were defective. The faults began with the fact that it would take a long time for the steam generated to be hot enough to start the vehicle. Also, it only had two small water tanks, each one only good for about twenty miles. Finally, the car was *prone to kerosene fires!*

Due to vehicle defects, Walker's former partner, Lorenzo Barber, and his **Mobile Motor Car Company** went into receivership. However, the **Locomobile Company of America** switched to gasoline engines in their vehicles. In 1901, the *Locomobile* was the top-selling auto in the United States. The company continued its success until 1922, when it joined the **Durant Motor Company**, a consortium put together by **General Motors** founder Billy Durant. It was the "top of the line" for the consortium, but it was not one of the surviving companies when that consortium broke up.

An early Locomobile and a later model

The *Corbin* was manufactured from 1905 to 1912 in New Britain, Connecticut, by the **Corbin Motor Vehicle Company.** Founded by home company **American Hardware Corporation**, they purchased the **Bristol Motor Company** before starting the **Corbin Motor Vehicle Company** and manufacturing the *Corbin*. The original version was a four-cylinder car with a ninety-four-inch wheelbase. It was named after Philip Corbin of the family that established the **Corbin Screw Corporation** and the **Corbin Cabinet Lock Company**. These were all divisions of the **American Hardware Corporation**.

1908 Corbin

From 1905 to 1908, the car's engines were air cooled, and from 1908 to 1910, both water- and air-cooled autos were offered. In 1910, only the water-cooled engines were provided. The cars generally sold for $2,000 to $3,000.

The car was known as "the full-jeweled *Corbin*" and enjoyed enormous success. Upon discontinuation of the brand in 1912, it was just said that it had "retired," as founder Philip Corbin had. The company produced about six hundred cars over the life of the brand.

In Middletown, Connecticut, the **Keating Motor Company** started in 1899 as the **Keating Wheel Company**, then changed to the **Keating Bicycle and Automobile Company,** and then, finally, the **R. M. Keating Motor Company.** Unfortunately, the company spent more time working on their name than manufacturing the electric, one-cylinder runabout vehicle they had announced was forthcoming.

In 1901, John W. Eisenhuth and the **Eisenhuth Horseless Vehicle Company** took over the **Keating Automobile Company** amid charges of blackmail and grand larceny. John W. Eisenhuth was arrested on at least two occasions, but this didn't stop the 1904 *Compound* from hitting the market.

The *Compound* was advertised as a light touring car selling for $1,400 in 1904. The design of John W. Eisenhuth, it had a three-cylinder, 20-horsepower gasoline engine on a chassis with a one-hundred-inch wheelbase.

The *Compound* auto was highly rated, especially for its fuel efficiency. It would probably have lasted beyond 1908 if the company had not scared off most of the buying public by originally overpricing the car at $6,000. They changed the company name to the **Eagle Motor Company** and lowered the auto's price to the $2,000 range, but the car never really caught on with the general public after the pricing snafu. In 1917, Eisenhuth was in Los Angeles, California, with a new, five-cylinder car called the *Poppy*. He died in 1918 before any cars could be manufactured.

CHAPTER 8:

THE DELAWARE CAR COMPANY OF THE DU PONTS

THE DU PONT FAMILY IS known as one of the wealthiest in the United States, famous for building a Wilmington, Delaware, factory that produced gunpowder. By the early 1900s, a descendant of Pierre du Pont, also named Pierre du Pont, headed up **General Motors** after Billy Durant was forced out.

A different Dupont family member, Paul du Pont, formed **Du Pont Motors, Inc.,** in 1919 after World War I restrictions were lifted. A luxury car, its motto was "as aristocratic as its name," which signaled who the car would be marketed to. It wasn't introduced at the New York Auto Show but rather at the classier 1919 International Salon at the New York Commodore Hotel.

The cars were built in a factory in Wilmington, Delaware, until 1922, when assembly of the car was moved to Moore, Pennsylvania. In 1925, it was moved back to Wilmington until 1932, when the company went into receivership. Never a large producing auto plant, the **DuPont** factory produced about 550 autos in all.

The first **DuPont** car was a four-cylinder model and sold for $2,600. The next model in 1925 was the six-cylinder *Model D*. The *Model E* came with a supercharged engine, but the most popular model was the one that followed.

Autos had come a long way from being called a horseless carriage and detested by everyone (except the rich) to becoming something that everyone wanted. A lot of early auto discontent was due to the people wealthy enough to be able to buy an auto like the **DuPont** models speeding through populated areas, scaring the horses *and* the people. In some areas, the residents would pelt the motorists with dirt clods (among other items).

1928 DuPont Model G

The *Model G* was the ultimate rich man's car and was introduced in 1929 with four forward speeds and a straight eight-cylinder engine. The *Model G Speedster* was guaranteed, with modifications, to go one hundred miles per hour! Celebrities who owned the *Model G* included Douglas Fairbanks (purchased for him by Mary Pickford), Jack Dempsey, and Will Rogers.

In 1930, Paul DuPont bought the **Indian Motorcycle Company** in Springfield, Massachusetts, and manufactured the car there for its remaining years. In the Great Depression year of 1932, they went into receivership, and Delaware's only car company to date became defunct.

CHAPTER 9:

FLORIDA CAR COMPANIES BUILT BOATS, TOO

THE **GAINESVILLE MOTOR COMPANY** WAS organized in 1906 to build automobiles, motor boats, and marine engines. Their goal was to make the first steel factory in Florida and build the autos there. They had reputedly leased land across from the Gainesville Post Office, but that was the last that was heard of them.

CHAPTER 10:

THE HANSON COMPANY OF GEORGIA

THE **HANSON MOTOR COMPANY** WAS a successful auto company. They sold an auto called the *Hanson* that had a six-cylinder Continental engine, available as the *Model 50*, and the larger *Model 66* in both open and closed models. Operating from 1917 to 1923, they sold hundreds of cars throughout the South and other areas of the United States.

The founder of the **Hanson Motor Company** was George Hanson, who in 1907 started selling

Hanson Six advertisement accentuating the South.

Franklin motor cars from his bicycle shop in Griffith, Georgia. He tore apart a *Packard* and developed his own vehicle. He had the car built in Detroit because it had lots of auto factories. The headquarters was in Atlanta, so the Hansons felt they could claim the car was "tested and approved in the South" and "made in Dixie."

Their first car, manufactured in 1918, was a five-passenger touring car. With the post-World War I recession, Hanson slashed prices. The "Little Six" model was sold for $995.00.

Hanson closed his factory in 1925 after producing thousands of autos. He went from manufacturing autos to manufacturing baby bottles.

Chapter 11:

The Electric Car Company That Never Happened in Hawaii

In 2010, the governor of Hawaii announced that Hawaii's first automotive venture would be a plant on Oahu for the **CT&T United Electric Car Company** of South Korea, who wanted to expand building their electric vehicles to the United States. They said they would employ four hundred people to manufacture ten thousand cars per year. They have advertised themselves as the world's largest producer of battery-powered electric cars. In 2014, the newly titled **CT&T South Korea** announced a new electric vehicle for the United States.

As of 2019, the vehicle has not yet appeared in the New World, and as of July 2017, the company admitted that its plans had stalled, and they were abandoning plans to build a plant in Pennsylvania. Company representatives announced that plants in South Carolina and Hawaii were still in the planning stages. A bad sign was when their website domain, www.ctntunited.com, went up for sale.

CHAPTER 12:

GOVE MOTOR COMPANY IS THE ONLY TRACE OF AN IDAHO CAR COMPANY

THE ONLY CAR COMPANY OF Idaho probably belongs on the scam list. In trade press releases of 1921, a reference is made to an auto named the *Gove* from Pocatello, Idaho, that was circulating a stock prospectus, looking for investors.

The *Gove* was reputedly built by H. E. Gove and the **Gove Motor Company** from Pocatello. One notice says that they are looking for factory space for an auto to be assembled in Detroit. They had a prototype that was built in Detroit or nearby Brighton. The company said they would also produce tractors and trucks. A 1½-ton, four-cylinder *Gove* truck, priced at $2,250, was slated for production. A *Gove* car was to be priced $100 less at $2,150. The car was to have a 110-inch wheelbase (or 114-inch, depending which source you were reading). A new plant in Pocatello was planned once enough stock had been sold. Not much more information ever materialized.

CHAPTER 13:

ILLINOIS CAR COMPANIES GLIDING THROUGH HISTORY

The Glide

A smooth-running, swift-going car. Originator of the new mode of travel:

GLIDING.

The car that satisfies the expectations stimulated by reading our competitors' advertisements.

Investigate before you buy elsewhere.

Travels the city or country as fast as it is desirable to go, at your own pleasure. It is comfortable, durable, reliable.

Price is Right.   **Selling Fast.**

Our catalogue tells it all.
Buy a Glide to save money and get service.

The Bartholomew Co.

Peoria, Ill.   110 Bradley Ave.

IN 1903, THE *GLIDEMOBILE* FROM Peoria, Illinois, became available to the public. The next year, it shortened its name to just the *Glide*. Developed by peanut and coffee roaster maker J. B. Bartholomew, the company also sold motorized popcorn poppers and peanut roasters. The *Glide* started off as a one-cylinder vehicle but quickly moved up to having two cylinders. Soon, the company was manufacturing four- and six-cylinder vehicles. By 1916, *all* their vehicles had six-cylinder engines. Contrary to most auto companies, Bartholomew kept his auto company intentionally small, confining production to a couple hundred annually. He proclaimed that he was happy, making enough to live well in Peoria.

As happened so often in the early independent auto companies, J. B.'s brother O. Y. Bartholomew worked for the company also, creating advertising. Some of his slogans were "Ride in a *Glide* and then decide!"; "As good as the best and better than the rest"; and "The chauffeur's choice, the owner's pride, the dealer's opportunity." The dealers eventually refused to take the small orders that the **Glide** company would provide and went to other automakers to stock autos for their customers. The *Glide* autos stopped being manufactured in 1920, although the Bartholomew brothers stayed invested in farm machinery and truck manufacturing.

The first prototype of the *Woods Mobilette* was in 1910, but it wasn't until 1914 that it got manufactured in Harvey, Illinois. The two-seater cyclecar had a four-cylinder engine. It sold for $380, with an optional windshield and top for $15 more. In 1916, electric lights were an option and were announced as standard for 1917. However, 1916 was the last year that the car was manufactured.

1915 Woods Mobilette

The **Henney Motor Car Company** of Freeport, Illinois, was mostly known for the manufacturing of funeral hearses, but they also manufactured about fifty custom-built sports touring cars from 1921 to 1931. They also manufactured taxicabs and a few closed sedans and limousines, powered by Continental six-cylinder engines or Lycoming eight-cylinder engines.

They started out making horse carriages in 1868 in Cedarville, Illinois, when they saw a need for funeral hearses. Their first hearses were elaborate carriages, driven by horses. When Cedarville was bypassed by the railroad, the business relocated to Freeport in 1878.

As time passed, the business came under the purview of father Jacob; then his son, John; and, finally, his grandson, John Jr. On their own, they produced over four thousand vehicles. In partnership with Packard, they built hundreds more.

The **Henney Motor Car Company** also made bodies for many diverse autos, including *Pierce-Arrow, Pontiac, Austin, Ford, Essex,*

Lincoln, Velie, and *Oldsmobile.* When they stopped making their own autos in 1931, they continued making seven-passenger sedans, limousines, and concept cars for the **Packard Motor Company**.

A 1940 Henning-Packard funeral hearse

In 1955, the company produced a prototype called the *Henney Kilowatt,* a vehicle called the first modern electric car. It had a range of fifty to sixty miles for each charge.

Henney-Packard ambulance and Henney touring sedan

The **Yellow Cab Manufacturing Company** of Chicago, Illinois, manufactured cars named the *Ambassador* and the *Hertz* from 1921 to 1928, when they sold out to **General Motors.** Hertz became the rental-car agency, was repurchased by original owner John Hertz in 1953, and was then renamed the **Hertz Corporation.** They no longer manufacture autos; instead, they rent them out.

———————————

In 1909, the initials of company founders T. L. Fauntleroy, H. R. Averill, and E. H. Lowe were used to form the name of the company and the auto they manufactured. **FAL Automobile Company** produced the *Fal*, sometimes called the *Falcar*. Although obviously not the most marketable name, *Fal* was an improvement over the first one: "*The Car With No Name.*" The car was built in the former Chicago factory of the **Reliable-Dayton Motor Company** following a search for a factory that led to Champaign and a deal that was reneged on, then to Waukegan, Illinois, and then Kenosha, Wisconsin, both of which also fizzled out. They finally ended up back in Chicago.

The car had a four-cylinder, L-Head Buda engine and was offered in Toy Tonneau, Roadster, "Speed Car," Touring, and Town Car styles. It was available in three sizes—a two-seater, a four-seater, and a seven-seater—and it had a three-speed transmission.

In August 1911, the founders of the company were hit by a lawsuit which forced them into receivership. Most of the petitioners causing the lawsuit were the farmers in the area who had purchased stock. They claimed that the assets of the company were being "neglected and dissipated."

In early 1912, the company was taken over by C. J. Marhoefer, who in 1913 introduced the *Grayhound*, a two-person roadster with "sleeping provisions." The front seat cushion could be pushed forward and used for a pillow, while the rear deck somehow provided "eight feet of sleeping comfort." The car didn't catch on, and probably, the "Cyclops look" of having only one headlight in the front didn't help. The company went into receivership. A. O. Dunk of the **Puritan Machine Company** of Detroit purchased the remnants of the company at an auction.

Ed Overholt of Galesburg, Illinois, started the **Overholt Company** in 1909 to produce his car, the *Illinois*. Considered a well-made car, it was small, with an eighty-six-inch wheelbase and a two-cylinder engine in a 12-horsepower runabout. In 1910, Overholt came out with a four-cylinder, 40-horsepower Reeves engine in his 110-inch wheelbase auto with a price tag of $1,350. For the next few years, his cars got faster, and in 1912, Overholt offered *two* four-cylinder engines in its latest car.

After manufacturing a little over one hundred autos, in 1912, he called it quits on producing autos. He bought the assets of **Reeves Pulley Company** and manufactured auto horns, spring-driven starters used in the *Ford Model T*, and engines. When his factory burned down in 1912, he retired to California.

The **Triumph Motor Car Company** of Chicago had a somewhat successful run, lasting five years, from 1907 to 1912. They sold the four-cylinder, 30-horsepower *Triumph "Self-Starting Car."* It had a four-speed transmission with a shaft drive and was available in runabout, touring, and limousine models.

The principals of the company were the Christopher brothers, Eric, R. B., and M. E., along with John Behrens. They sold out after just several months to Vincent Bendix and O. M. Dulauney, two former **Holsman** employees.

The price of the vehicle went down during the life of the car from $3,500 in 1907 to $2,250 in 1912, but they didn't have enough profit to keep going after 1912. Bendix went on to be associated with the *Duplex* and *Bendix* autos.

The **Holsman Automobile Company** made high-wheeler autos from 1903 to 1910 in Chicago and from 1910 to 1911 in Plano, Illinois. They had two-cylinder, 10- and 12-horsepower engine cars and four-cylinder, 26-horsepower autos. They were at first driven by 7/8-inch manila rope until 1905, then steel cable from 1905 to 1909, and then chains in 1909 and 1910.

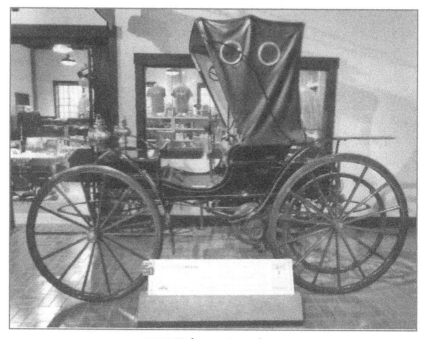

1909 Holsman Runabout

The *Holsman* not only sold well in the Midwest, but they also made international sales in India and Australia. They called themselves the "oldest motor-buggy makers in America" and featured the high, buggy-like wheels that won hill-climbing events in the car's advertising. Once better roads came about, the high-wheeler autos faded out.

The **Moline Plow Company** was founded in 1870 in Moline, Illinois, and combined with smaller firms in the area to form a thriving farm implement and buggy company until World War I. In 1916, the company announced that since the farm and buggy business was going downward, they were going to begin manufacturing automobiles as the **Stephens Motor Car Company.**

Moline Plow Company, Moline, Illinois

In 1917, the first car appeared on the market, the *Stephens Salient Six,* named after company president, G. A. Stephens, and his father, who founded the original company. It was a six-cylinder, 35-horsepower car and came in touring and roadster models.

In 1919, John Willys of the **Willys-Overland Motor Company** purchased a controlling interest in the **Moline Plow Company,** the parent company of the **Stephens Motor Car Company.** In the early 1920s, production levels topped at three thousand cars manufactured per year. In 1923, to combat falling sales, the car was restyled, but by 1924, the **Moline Plow Company** discontinued autos and went back to making farm implements.

The Coey of Chicago

The **Coey-Mitchell Automobile Company** of Chicago, Illinois, built a car called the *Coey* from 1913 to 1917. It was available in two models, the *Bear,* a cyclecar, and the *Flyer,* a full-sized model with a six-cylinder engine. The *Coey Flyer* came with a 124-inch wheelbase, a three-speed Harry Stutz transmission, and a Presto starter. Options included a top, front glass, curtains, five lamps, tools, an air pump, and a jack. The *Coey Bear* was more of a cyclecar and came out in 1915. It was a two-seater roadster with a narrow, ninety-six-inch wheelbase. World War I shortages hurt the company, and they switched to producing commercial vehicles. They were purchased by the **Wonder Motor Truck Company** in 1916.

The **Comet Automobile Company** built the *Comet* in Decatur, Illinois, from 1917 to 1922. It was a car using both Lewis and Continental six-cylinder engines, which sold in "small but consistent" numbers. Both open and closed models were available. The brainchild of George Jagers, he built the first prototypes of his *Comet* auto by first acquiring controlling interest in the **Racine Manufacturing Company**, a toy and novelty business that had seen better days.

Comet auto

Comet radiator badge

Taking his first *Comet* prototype to the Chicago Auto Show in 1917, he was encouraged to begin production, which he did in August of that year as the **Comet Motor Company**. He acquired a factory in downtown Decatur, Illinois, and assembled his six-cylinder, 50-horsepower auto with a 123-inch wheelbase.

As the car received orders, it would assemble autos. Its largest production year was 1920, when the number of cars assembled was over six hundred. It seemed like its motto, "The Comet is Here to Stay," would hold true. They even received an order for forty cars from Antwerp, Belgium. However, this only served to make them assume that more overseas orders were coming—orders that never came.

As sales dwindled and a bond sale failed, the company went into voluntary receivership in 1921. They released a four-cylinder auto that only cost $1,000, but it was too late. The *Comet* stopped being manufactured in 1922 and, failing to find anyone to buy the whole company, auctioned off the remaining inventory in 1923.

Preston Tucker from Ypsilanti, Michigan, formed the **Tucker Corporation** to build the car he developed, the *Tucker*. Only forty-nine were assembled in a Chicago plant that was formerly used by **Dodge** to make aircraft engines. Accused of fraud and violation of Securities Exchange Commission regulations, Tucker lost a costly court battle but was vindicated in 1950. His plans to build a compact car in Brazil ended when he died in 1956.

The 1948 *Tucker Torpedo* had many amenities not generally available, including padded dashboards, a front-passenger crash compartment, a pop-out windscreen, and disc brakes. The car, with its iconic third headlight, has gone on to become one of the most prized collector autos with prices of the remaining autos selling for over a million dollars apiece.

The 1948 Tucker Torpedo and the 1927 Velie Model 50

The **Velie Motor Vehicle Company** of Moline, Illinois, started with William Lamb Velie making carriages in 1902. In 1909, the first auto, called the *Velie*, was launched, with a 35 horsepower, four-cylinder engine. The company had backing from the **John Deere Plow Company** (Velie was married to Emma Deere, daughter of John Deere) and sold the cars through John Deere dealerships.In 1917, the company switched to using only six-cylinder engines and started constructing their own engines in 1922. In 1927, they continued making their six-cylinder engines and also offered a straight eight-cylinder Lycoming engine. Although the company went under in 1928, the name survives in Louisiana, where the town and street of Velie were named after the auto.

1927 Velie auto Model 50

W. E. Clark of Moline, Illinois, built his first car in 1897, a one-cylinder, air-cooled vehicle, but never marketed it. He likewise had a 1901 car that never hit the market. A brief foray in 1903 was a car called the *Blackhawk*. This car was unsuccessful, and Clark spent the next two years looking for financing, which he found in 1905 from the **John Deere Company.** The well-known farm equipment manufacturer bought out the remains of the **Clarkmobile Company** of Lansing, Michigan, for their auto venture.

The car company was established under the name the **Deere-Clark Motor Company** in Moline in 1906. The car they manufactured as the 1906 *Deere-Clark* was a water-cooled, four-cylinder, 30-horsepower vehicle with a one-hundred-inch wheelbase, offered in touring and limo models. In 1907, they added a *Gentleman's Roadster* model and shortened the company name to just **Deere**. Their motto was "A Car Designed to Satisfy." However, the **John Deere Company** soured on this venture after only two hundred units were produced. The company was reorganized as the **Midland Motor Car Company.**

The name of the auto was changed to the *Midland* with the 1908 models. It was advertised as "The Motoring Sensation of 1908" by new investor Charles H. Pope. *Midland*s had four-cylinder, 30-horsepower engines and were available in five- to seven-passenger touring models. In 1909, a roadster model was added. By 1910, the horsepower was upped to 50 for some models. In 1912, the Toy Tonneau and Colonial Coupe models were added as well as a six-cylinder engine option and a two-passenger roadster model. The autos sold in the $2,000 price range, but after selling over 2,500 autos, the company went into receivership in 1913 with over $450,000 of debt. Accounting was slipshod, and between forty and fifty cars disappeared without serial numbers. Payroll discrepancies and more led to the assets of the company being sold in 1914.

Chapter 14:

Indiana's Infinite Number of Car Companies

NDIANAPOLIS, INDIANA, WAS THE LOCATION of many successful auto companies and famous auto marques, including **Haynes, Marmon, Cole, Stutz, Crow-Elkhart, Marion, and Lafayette**. Therefore, Indianapolis was almost the Motor City. Besides Indianapolis, the rest of Indiana had some major auto companies. These included—from Auburn and Connersville, Indiana—the **Auburn, Cord,** and **Duesenberg**. Other big Indiana car cities were Elkhart, Anderson, South Bend, and Kokomo, both of which would have been in Michigan if not for the circumstances of the Toledo War.

For Indiana, the best place to start is with Elwood Haynes, who was the inventor of one of America's first automobiles in Kokomo, Indiana, in 1894. Since his gasoline-powered vehicle seemed marketable, in 1906, he partnered with the Apperson Brothers, Elmer and Edgar, to build it, and by 1916, the **Haynes-Apperson Motor Company** had made and sold over seven thousand cars. The cars were known for their reliability in traveling long distances.

Elwood Haynes

1894 Haynes-Apperson

In 1902, Haynes split from the Apperson brothers. The Appersons continued to produce autos and manufactured some of the first four-cylinder and six-cylinder cars until the company liquidated in 1926. Haynes started the **Haynes Automobile Company**. In 1905, Haynes's company released their most renowned car, the *Haynes Model L,* which could carry four passengers in the three-speed auto. It could do thirty-five miles per hour, and over 4,300 were sold.

The Haynes factory

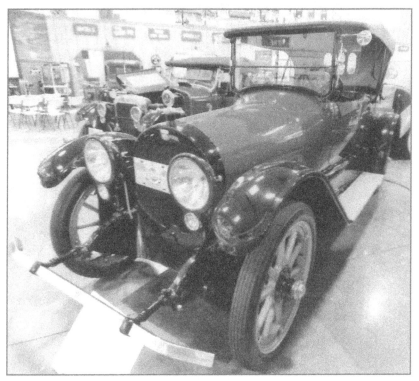

1917 Haynes

Haynes expanded the line in 1905 because of increasing sales. At this time, their annual production was 650 cars per year. In 1910, they became the first auto company to include a windshield, roof, speedometer, and headlights as standard equipment on every car they sold. They enjoyed prosperity until 1911, when a factory fire slowed them down. In 1913, they began production again. To gain publicity, Elwood Haynes did a national tour, appearing in many places billed as "Father of the Automobile." This title caused consternation from the Duryea Brothers, who also claimed the title. Haynes received a patent on *Stellite*, his version of stainless steel, in 1912.

1916 Apperson and logo

From 1909 to 1924, the **Crow-Elkhart Motor Company** manufactured autos in Elkhart, Indiana. It was founded by Martin Crow, and the cars are renowned as having one of the first auto self-starters. The first auto of the **Crow-Elkhart Motor Company** was the *Martin E. Crow,* a passenger touring car. The 1918 models sold for $935. In 1922, the company went into receivership, and the assets sold in 1923 for $78,000.

Crow-Elkhart and Crow-Elkhart advertisement;
Crow-Elkhart badge

The Crow-Elkhart

The **Lafayette Motors Company** of Mars Hill, Indiana, manufactured a large, luxurious touring car with a V-8, 100-horsepower engine. They operated from 1920 to 1924 and then were taken over by the **Nash Motor Company. Nash** kept the *Lafayette* marque as a luxury line for a little while and then dropped it until 1934, when they used it as the brand name for its *least* expensive vehicle.

1924 Lafayette

The **Marion Motor Company** of Indianapolis manufactured the *Marion* from 1904 to 1915. Like the *Premier*, they had four-cylinder, 16-horsepower Reeves engines and, later, water-cooled, six-cylinder Continental engines. They had large, four- and six-cylinder cars, especially in the 1912 to 1914 period, but their most renowned was the V-12 they built in 1908.

The motto for the company was "The Car That Has Set Men to Thinking." In 1914, J. J. Handey purchased the company, moved production to Jackson, Michigan, and changed the name of the car to *Marion-Handey*.

1905 Marion of Indianapolis

1912 Marion Roadster

The 1919 Cole Auto and the Cole radiator badge

The United States had over a thousand auto companies that produced cars for at least a year, and Indiana was probably good for at least a fifth of those. While you didn't have to be one of the top automobile companies to make money, usually, the small, independent companies needed to stay afloat for at least three years before turning a profit. However, if you thought that no one other than the Big Three made money, you are wrong. Many of the smaller companies, such as the **Cole Motor Car Company**, lasted for many years and made their owners millionaires. Owner J. J. Cole was one, and when he wanted to retire, he wouldn't sell his company to anyone else. He didn't want anyone operating HIS car company, so he liquidated it. He didn't want to risk having some-one run it who would not have the integrity or honesty that he did since it was his name on the car.

J. J. Cole was the inventor who forgot to put brakes in his first car and had to drive around and around the Indianapolis Soldier and Sailor Monument until he ran out of gas so the car would stop. In spite of what would be the shortsightedness of its inventor, the *Cole* was considered to be a well-built, quality luxury car. Using one of the first V-8 engines, the **Cole Motor Company** built cars in Indianapolis from 1908 to 1925.

In 1903, J. J. Cole worked with his son on an auto in Rockford, Illinois, to be called the *Rockford*. Their workshop was the back of the shop they owned and where they sold wheels, motor bikes, and lawn mowers and did repair work. When they were unsuccessful with their auto, they opened a *Rambler* automobile dealership.

Not giving up, in 1904, Cole bought the **Gates-Osborne Carriage Company** and renamed it the **Cole Carriage Company**. He built his first successful auto here in a two-cylinder, high-wheeler motor buggy style.

In 1909, the auto was changed to a more conventional, albeit smaller, two-cylinder runabout style and sold in the $750 range. About one hundred were sold. More successful was the larger model released in 1910, which had a four-cylinder engine and was priced at $1,500. This model was available in four body styles, including a runabout called the *Cole Flyer*. The company sold 783 of the new models and even won a few races!

In 1911, a four-door model was added as well as a longer version for $1,650. In 1912, similar cars were added, selling for $2,500 and $3,000, bringing the *Cole* autos into the luxury class. In 1915, one year after *Cadillac* and *Cunningham* debuted theirs, *Cole* brought out a V-8 engine on *all* of their models, getting rid of all the four- and six-cylinder models. The *Cole* brand successfully continued for another decade, finally calling it quits in 1925 as Joseph J. Cole liquidated the company. He died on August 8, 1925, of an infection shortly before the liquidation was complete.

Cole Motor Company plant

The **Pratt-Elkhart Motor Company** grew out of the **Elkhart Carriage and Harness Manufacturing Company** of Elkhart, Indiana, in 1909. Brothers William and George Pratt had great success advertising their carriage business in mail-order catalogs.

They made their first auto in 1906, and they were not happy with its performance. It took them until 1909 until they felt they had a marketable vehicle. The one they banked on was a four-cylinder, 35-horsepower touring car. However, the mailing lists they had organized for the carriage business didn't pan out for autos because the autos the Pratt brothers sold were more expensive than the carriage that customers could afford.

The Pratt brothers decided to focus more on finding dealers to sell their cars, since they couldn't bring themselves to lower the price on their cars by lessening the quality. In 1911, they became the **Pratt Motor Car Company,** and the brothers offered open and closed touring cars, roadsters, and limo models.

In 1914, the Pratts added six- and eight-cylinder engines in a single model. In all, a little over a thousand *Pratts* were manufactured during the life of the car. In 1915, the company did a turnaround and went back to selling carriages. By 1917, they concentrated on producing ambulance bodies as World War I production geared up.

The Pratts also worked on the *Elcar* of Elkhart, Indiana, as the **Elkhart Carriage and Motor Car Company.**

The *Elcar* was first produced in 1916 as a four-cylinder, 35-horsepower vehicle available in runabout and touring styles and selling for only $795. It was advertised as the "Car for the Many." In 1918, a six-cylinder model was added. When the Pratt brothers retired and sold their shares to **Auburn Motor Company** executives in 1921, the name was changed to the **Elcar Motor Company.**

Business was consistent, and the company added a line of taxicabs when winning a contract to manufacture one thousand cabs in late 1922 for the **Diamond Taxicab Company** in New York City. Other marques they manufactured were the *Martel,* the *Royal Martel,* and the *Elfin.*

For the regular *Elcars,* a straight Lycoming eight-cylinder engine was offered in a car in 1925 as well as Lockheed hydraulic brakes. The four-cylinder cars were eliminated in 1926. The horsepower of the cars increased from 65 until 1929, when a 140-horsepower *Elcar* hit the market.

By 1930, the Great Depression was hitting the auto market hard. The company went into receivership and then was bankrupt in 1931. From 1916 to 1931, over nine thousand *Elcars* were built in Elkhart.

––––––––––––

The *National* was built in Indianapolis and was a seven-passenger touring car. It had a four-cylinder T-Head engine with a horsepower of 50. The price tag was $4,000.

The National of Indiana

The *National* auto started in 1900 as an electric vehicle and added a gasoline vehicle in 1903. In 1905, the electric vehicle was discontinued in favor of the gas car, and the **National Automobile & Electric Vehicle Company** became the **National Motor Vehicle Company.** In 1906, they began offering a six-cylinder engine. A distinctive, round radiator characterized the *National* autos. From 1909 to 1912, the *National* won several car races, including the Indianapolis 500 in 1912. In 1915, a twelve-cylinder engine was offered in place of the six-cylinder on some models.

In 1922, the **National Motor Vehicle Company** merged with the company that made Kentucky's *Dixie Flyer, the* **Dixie Motor Car Company**, and the **Jackson Automobile Company**, which made the *Jackson.* Each of these cars was offered from 1922 to 1924 under the *National* name.

The Stutz Motor Car Company began in 1911 in Indianapolis, Indiana. Its founder, Harry C. Stutz, first built an auto in his native Dayton, Ohio, out of his small machine shop. Moving to

Indianapolis, he worked with the **Empire Automobile Company** and others before forming his own company, naming it the **Ideal Car Company** and locating in a factory at 1002 North Capitol Ave. in Indianapolis. In 1913, the company was renamed the **Stutz Motor Car Company.**

The *Stutz* cars were high-performance, luxurious cars available in mostly roadster styles; the most famous was the *Bearcat*. Most models were available with either four- or six-cylinder engines and sold in the $2,000 range. By 1925, the four-cylinder models were phased out. The *Stutz Bearcat* won many races and was a favorite car for both amateurs and professional race car drivers.

Stutz ad

1935 Stutz Bearcat

Stutz Bearcats

From 1928 to 1930, the company put out a smaller, less expensive version of the *Bearcat* called the *Blackhawk*. It was discontinued upon the ascent of the Great Depression. The **Stutz Motor Car Company** was declared insolvent in 1937 and liquidated in 1939.

Connersville, Indiana, was the home of the **McFarlan Motor Car Company**, a successful, prosperous, eighteen-year car company. Like so many others, the company got their start as the **McFarlan Carriage Company**, one of the most successful horse carriage manufacturers in Indiana.

The *McFarlan* was a luxurious, six-cylinder vehicle available in runabout, touring, and torpedo models in its first year, 1910. John McFarlan was the family patriarch, and Harry McFarlan was the upstart son who wanted to manufacture autos. Sadly, John died two weeks before the release of the first *McFarlan* in 1909. The company made both carriages and cars until 1913, when the carriages were discontinued.

The *McFarlan* production was never more than two hundred

autos built per year, but the ones they built they had no trouble selling, even at the $6,000 price tag attached to the later models. Only about 3,800 of the cars were built in the eighteen-year history of the marque.

McFarlan Carriage Company,
*forerunner of **McFarlan Motor Company***

1925 McFARLANE, owned by Jack Dempsey on display at Movie World, Buena Park, California.

In 1921, the company built its own engine after experimenting with using various other companies' engines over the years. The engine they developed was called the Twin Valve Six, which used eighteen spark plugs and had triple ignition. It was put into the *Knickerbocker Cabriolet,* an auto which sold for $9,000 with 120 horsepower. In 1926, a straight eight-cylinder car was an option.

In 1928, the car was restyled, but this was the last year for the stylish marque, as Harry McFarlan died. When Harry died, his car died with him, and the company was sold to Errett Cord, who wanted the factory to use for building his latest car, the *Duesenberg.*

Charles Eckhart and his sons, Frank and Morris, operated the **Eckhart Carriage Company** in Auburn, Indiana. In 1900, Frank bought a *Curved Dash Oldsmobile* while in Lansing making sales calls. Driving it home, his father and brother helped him put together their own one-cylinder, chain-driven auto they named after their city, *Auburn,* and their company became the **Auburn Automobile Company**. The 1904 and 1905 *Auburns* had two cylinders; in 1909, they added a four-cylinder model and, in 1912, a six-cylinder model was released. Touring and Tonneau models were added to the line as production went up each year.

1912 Auburn Model 30-L Roadster

The Eckharts had produced over twenty-three thousand *Auburns* by 1919, when World War I was making it hard to get parts. They sold the whole company to a consortium that included chewing gum mogul William Wrigley.

The infusion of capital made a big difference in the auto. The 1919 *Auburn Beauty Six* was the beginning of a larger, more luxurious car that made the *Auburn* the choice of many tycoons in the Roaring Twenties. Errett L. Cord became the chief engineer in 1924, and the next year, 1925, a lavish, eight-cylinder *Auburn* with two-tone colors and lots of nickel plating was introduced. The new model proved so popular that the company doubled their sales three years in a row. The line was broadened to include phaetons, touring cars, roadsters, runabouts, and town coupe models.

In the meantime, Fred and August Duesenberg were launching their self-named car, the *Duesenberg,* in 1920 in Indianapolis. The cars of the **Duesenberg Motor Company** were at first known mainly as race cars, but they soon began producing luxury passenger cars as well.

The Duesenbergs started, like so many other car guys, in the bicycle business. From there, they manufactured car and marine engines. Going into auto manufacturing in 1920, they produced some of America's most iconic cars.

Fred was an engineer's engineer, but neither brother was great at business. Going into receivership in 1929, the company was purchased by Errett Cord of the **Auburn Automobile Company**. *Duesenberg*s became more widely available as production was moved to Connersville, Indiana, in 1929, to the old **McFarlan** factories. The same year, a new luxury car, the *Cord,* was introduced.

1932 Auburn 12-161 Convertible Coupe

With just a little dip in sales from the Great Depression in 1930, by 1931, they had recouped and continued making luxury cars. Errett Cord consolidated the holdings into the **Cord Corporation.** The firm made some of the most highly regarded cars in auto history, including the *Model J Duesenberg* and the front-wheel drive 1929 *Cord.* Fred Duesenberg stayed on as chief engineer for the company until his untimely death in an auto accident in 1932. His brother Augie took over as chief engineer.

1934 Duesenberg Model SJ

In 1936, the models were the same as the year before as Errett Cord was investigated by the IRS and the SEC. By 1937, E. Cord had sold his stock and moved to Beverly Hills, California, and the company that produced America's most elegant cars went out of business.

In Terre Haute, Indiana, the **Overland Auto Works Company** (named during a coffee break) manufactured the *Overland* from 1903 to 1912. Designed by Claude Cox, in 1905, Cox and the company, renamed the **Overland Motor Company**, relocated their factory to Indianapolis.

In 1912, the company was purchased by John Willys and renamed the **Willys-Overland Motor Company.** From 1912 to 1918, the company was number two in the United States in sales (behind **Ford**).

A 1912 Overland Runabout model and
a 1910 Overland Touring model

Willys-Overland Toledo HQ

John Willys kept the *Overland* name on the cars. During the Panic of 1907, when he couldn't find factory space, he assembled the *Overland* autos in a circus tent. He purchased the **Marion Auto Company** so he could use their Marion, Ohio, factory. **Willys-Overland** morphed into **Nash, American Motors,** and **Chrysler**.

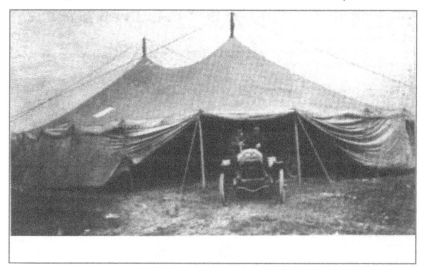

Overland tent factory

Joseph Frazer was a well-respected and valued employee when he left **Chrysler Motor Company** in 1939 to become president of the **Willys-Overland Motor Company.** He was even more respected when he arranged for the company to get the contract for the Jeep from the U.S. government. When Frazer left in 1944 to take charge of the **Graham-Paige** company, **Willys-Overland**'s annual sales were over $21 million. Joe Frazer partnered with industrialist Henry J. Kaiser (who was a tycoon that owned a ship-building company, a steel company, hospitals, Kaiser Aluminum, and more) and bought the controlling stock.

The company was renamed the **Kaiser-Frazer Company.** Well-known models were the sporty *Kaiser*s and the elegant *Frazer*s.

When Joe Frazer quit in 1948, the company was reorganized as the **Kaiser Motor Company.** In 1953, Henry Kaiser purchased the **Willys Company** and specialized in manufacturing *Jeeps* to the extent that the company name was renamed the **Kaiser-Jeep Corporation.** *Jeep* production kept the company going until 1970, when it was sold to **Nash-Hudson** successor **American Motors Company,** eventually becoming part of **Chrysler.**

1953 Kaiser

1949 Kaiser

The **George W. Davis Motor Car Company** of Richmond, Indiana, manufactured a four-cylinder touring model and then added a six-cylinder model. In 1927, they added an eight-cylinder car to their lineup. Manufacturing cars from 1908 to 1930, the autos had a wide variety of body styles. Two of the more noteworthy ones were the *Fleetaway* touring car and the sporty *Man' o' War* roadster.

The **Automotive Corporation of America** acquired the company in 1928 and continued manufacturing the *Davis Eight* and the *New York Six*, models which were available with the Parkmobile device, "which lifted the cars into tight parking places." After 1930, the factory's production changed to lawn mowers, aircraft, and power machinery.

In Richmond, Indiana, **Wayne Works** started making farm implements in the 1870s. In 1901, the president, Walter W. Schultz, asked employee Jack St. John if he could put together a car because "*everyone else is*." He did put together one, a two-cylinder, air-cooled touring car, which was quickly sold. The next year, they repeated the process, building one car and then selling it. Finally, in 1904, they felt they were ready to begin manufacturing more cars.

The *Richmond*, as the car was named, was a four-cylinder vehicle available in roadster and touring car models. The cars were known for their reliability and hill climbing. In 1910, the company switched from air-cooled engines to water-cooled. In 1914, they brought out a six-cylinder model.

In 1917, after manufacturing over 1,300 cars called *Richmond* in sixteen years, the auto division of **Wayne Works** was closed. The company continued making farm implements after their auto business closed.

The **Premier Motor Manufacturing Company** of Indianapolis started with George Weidely manufacturing and selling his first auto, a water-cooled motor buggy in 1902. In 1903, he joined with Harold O. Smith and produced a vehicle with a four-cylinder, 16-horsepower engine with overhead valves, sliding-gear transmission, and friction drive.

1916 and 1909 Premier

The 1909 Premier and the 1916 Premier

In 1903, they manufactured the four-cylinder *Premier* with an air-cooled engine. They assembled the six-passenger touring cars until 1925, a notable twenty-two years. The touring model of the

1919–1920 season used a lever on the steering wheel to control the electric transmission system. This was called the Cutler-Hammer magnetic gear shift. The *Premier* auto was designed by G. A. Weidely, and they had large, four- and six-cylinder autos of 24 to 28 horsepower in 1907; from 1913, only six-cylinder engines were used.

In 1902, Howard and Walter Marmon built their first auto in the factory of the **Nordyke and Marmon Company**, makers of flour-milling equipment in Indianapolis. Howard was the designer, while Walter handled the business side. The first *Marmon* engine was a V-4, air-cooled, overhead valve engine designed by Howard. The 1904 *Model A Marmon* was a touring car with a V-4, 20-horsepower engine; the 1905 model had 24 horsepower. It was advertised as "A Mechanical Masterpiece."

Roosevelt logo and 1924 Marmon

In 1907, a *Marmon* with a V-8 engine and 75 horsepower was released in a seven-passenger touring car selling for $5,000. Although most customers were scared off by the price, they came back to take a look when the *Marmon Wasp* won the first Indianapolis 500 car race in 1911.

By 1909, most of the air-cooled engines of the early cars were changed to water-cooled engines. The company kept producing larger cars (90–123-inch wheelbases) and kept on winning races and endurance contests through the decade of the 1910s. They were manufacturing suburban, touring, and roadster models, with larger engines of 80 horsepower. The first rearview mirror was on a *Marmon*.

Marmon Motor Car Company,
Indianapolis, Indiana

With production of the *Marmon*s reaching its highest numbers yet at 4,500 in 1926, the flour-milling business, which had supported the auto company through its lean years, was sold to **Allis-Chalmers** to allow more room in the factory to build autos. The company name was changed to the **Marmon Motor Car Company**.

In 1927, the *Little Marmon* was released. A success for the restructured company, it was introduced due to the new management's desire for a lower-priced car. The car increased its production from ten thousand to twenty-two thousand in less than two years.

In 1929, the popular *Roosevelt* model was released (named after Teddy, not FDR), and high production and sales of over twenty thousand vehicles ensued. Everything was going well until 1930, when the Great Depression started. Production plummeted from the thousands to just *eighty-six* in 1933, the year the company went into receivership. Too little too late, Howard was getting ready to introduce a sixteen-cylinder, 200-horsepower auto.

Howard's other brother Walter made trucks with his partner Arthur Herrington as the **Marmon-Herrington Motor Company** until the 1960s.

The 1929 Roosevelt and the Marmon-Herrington logo

CHAPTER 15:

THE SPAULDING, MASON, MAYTAG, AND MORE IN IOWA

THE *SPAULDING* WAS A FAMOUS Iowa car company—the one that manufactured the *Spaulding Sleeper Car,* which had a foldout bed in the back, complete with *a reading light!* The **Spaulding Manufacturing Company** was from Grinnell, Iowa, and, like many automakers, started out making carriages. From 1910 to 1916, they successfully made *both* carriages and cars. The company founder, H. W. Spaulding, began making carriages in 1876 and was doing over $1.5 million of annual sales by 1910 when he decided to enter the automobile business.

Henry W. Spaulding was born in Vermont but moved to Grinnell, Iowa, in 1876, where he opened a blacksmith shop and began making carriages on the side. He served as a city councilman and two terms as mayor of Grinnell before getting elected twice to the state legislature.

Spaulding had a setback in 1882 when a tornado completely destroyed his factory, but he bounced back, and his peak production rose to over twenty thousand carriages a year.

As plans were made to begin manufacturing motor cars, a third floor was built on to the brick **Spaulding** factory in 1909. In 1910, the *Spaulding 30* automobile came out and was available in two styles, a touring car and pony tonneau. Both had 30-horsepower, four-cylinder engines. In 1911, the *Spaulding 40* came out. It had a 40-horsepower engine, hence the "40." Because Iowa had very bad roads, the *Spaulding* autos were built with especially good shocks and spring suspension. The *Spaulding* models all had long, flat, elliptic front springs. One of the company's slogans was "You Pay a Little More, But You Pay Less Often."

The *Spaulding* was one of the nation's strongest "regional brands." They had dealers all over Iowa, including dealerships in Grinnell, Des Moines, Hampton, Muscatine, New Sharon, Spirit Lake, Ottumwa, and Tama. Outside Iowa were dealerships in Los Angeles and San Diego in California, Fort Worth in Texas, and Omaha in Nebraska.

Demonstrating the hill-climbing abilities of the 1913 Spaulding auto

An attribute of the *Spaulding* was its ability to plow through snow. Another was its climbing ability, as demonstrated by Ernest Spaulding when he took a 1913 *Spaulding* up the steps of the Iowa State Capitol Building. The *Spaulding Sleeper* was unveiled at the 1913 Iowa State Fair in Des Moines, Iowa. It could be converted into a sleeping compartment (complete with reading lamp) in a manner of minutes.

The factory primarily turned out *Spaulding 40* touring cars from 1913 to 1915. In August 1915, they advertised a small 1916 *Spaulding 30* for only $750. This was the company's answer to being undercut in price by the **Saxon, Studebaker**, and **Ford**, but the price cutting was to no avail, and by 1916, the **Spaulding**

Manufacturing Company was out of the car business. In 1917, the company started manufacturing a very successful manure spreader—so successful, in fact, that the company could not build enough to keep up with the demand. They later made bodies for **Ford Motor Company**.

The Spaulding factory complex was put on the National Register of Historic Places in 1978. It also houses the Iowa Transportation Museum. In 2017, a seventy-seven-unit loft apartment complex opened in the former auto factory.

Spaulding factory in Grinnell, Iowa

Fred Duesenberg's first design job was the *Mason* for the **Mason Motor Car Company** of Des Moines, Iowa, from 1906 to 1910. The auto, called the *Mason,* was a five-seater, powered by a 24-horsepower, two-cylinder engine. In 1909, a choice of a two-seater or a four-seater was offered.

In 1910, the **Mason Motor Company** was purchased by the appliance manufacturer Maytag. The new company was named the **Maytag Mason Motor Company**, and its headquarters was in Waterloo, Iowa. A two- or four-cylinder car with either 16- or 30-horsepower engines were offered. The *Model H* roadster of 1911 had a boat-tailed body.

Mason Motor Company, later Maytag

By the end of 1911, the company exited the auto business to concentrate on their more successful washing machine business, still in business in the twenty-first century.

The **Adams Company** in Dubuque, Iowa, started out in 1883 as the **Roberts & Langworthy Iron Works** making "fine iron castings" of such items as grave benches and crosses. They were joined by Eugene and Herbert Adams, and eventually, as Langworthy and Roberts retired, the business became just the **Adams Company**.

In 1893, their plant at 57 South Main Street in Dubuque burned down, and they built a new factory on East Fourth Street. They diversified into making laundry stoves and other items. In 1895, Fay Oliver Farwell joined the firm as plant manager and began experimenting with internal combustion engines.

By 1898, Farwell had built a prototype auto. He built four more before the company was ready to put their auto, the *Adams-Farwell,* on the market in 1904. The first auto sold had a three-cylinder, 20-horsepower radial engine. In 1906, a five-cylinder, 45-horsepower model was added. The *Adams-Farwell* autos had many distinctive features, but the most memorable was the circular device which allowed the car to be steered from the front seat or the *back* seat!

Another innovation was a four-speed transmission that involved *two* clutches in 1908. The bodies were constructed by the Connolly Carriage & Buggy Co. Usually about twenty-five cars a year were manufactured, with 1910 yielding about fifty. The entire marque production was about two hundred autos, with a number of them going to the Dubuque Police Department.

The auto company had a good run, but after 1912 (some sources say 1913), they discontinued making autos. The **Adams Company** is still in business today as a private corporation manufacturing custom gears and shafts.

A 1906 Adams-Farwell auto

Chapter 16:

Alias Smith and Jones Car Companies of Kansas

WHEN THEY FORMED IN 1900, the **Smith Automobile Company** of Topeka, Kansas, was one of the nation's first car companies. They were also one of the few automobile companies that made almost all their own parts, including transmissions, engines, and steering gear. They built their frames out of wood, which was customary in 1900.

1902 Smith of Topeka's Kansas Historical Society

They manufactured the autos named the *Smith, Great Smith,* and *Veracity.* Their first auto was the *Veracity,* which had a two-cylinder, water-cooled engine. From there, the *Smith* autos had larger, four-cylinder engines, while the *Great Smith* had the largest engine, with six cylinders.

As so many car companies were, the **Smith Automobile Company** was started by two brothers, Clement and Anton Smith. The Smith brothers (not to be confused with the cough drop Smith Brothers) started their manufacturing careers as makers of trusses for hernia patients. They went on to manufacture artificial limbs, concert grand harps, and bows and arrows before settling on autos in 1902.

The Great Smith automobile

Terry Stafford first started building successful, running automobiles in 1898, which he would drive around his hometown, Topeka, Kansas. Selling the rights to manufacture his car to the **Smith Motor Company,** he worked for them as their chief engineer until 1908. Stafford was granted patents for the **Smith** autos' steering gear, transmission, brake system, and drive train. Even though Stafford created the first *Smith* car, the Smith brothers released a company history and forgot to mention that Stafford went on to form the **Stafford Motor Company** in Kansas City, Kansas.

Terry Stafford first started building successful, running automobiles in 1898, which he would drive around his hometown, Topeka, Kansas. Selling the rights to manufacture his car to the **Smith Motor Company**, he worked for them as their chief engineer until 1908.

The **Smith Automobile Company** division of labor was that brother Anton handled the business, while Clement invented items for autos. He had three patents for carburetors and five other auto patents.

When *Smith* cars started winning races and setting speed records in 1908 and 1909, their sales grew to the point that almost half the autos in Topeka were from the **Smith Automobile Company.**

For advertising, the car factory had its own print shop. By 1904, they were producing well-designed, multicolor catalogs of their cars. A big advertising opportunity appeared for the company when in October 1908, a 1908 *Great Smith* became the first auto to conquer Pikes Peak in Colorado. Perhaps their most famous slogan was "The World's Greatest $2,500 Car," but they had many others, including "Smooth as a Bicycle" and "Capable of Hard Usage." The demand for **Smith** autos was said to be "Growing Like Alfalfa."

In 1903, the Smiths took the plunge and built a factory at 10th and Jefferson in Topeka. They went on to improve their vehicles. Innovation-wise, the **Smith** cars had telescopic steering wheels and semi-automatic transmissions. The *Great Smith* had an *icebox* on the running board!

The **Smith Automobile Company** had about 150 employees, building around one hundred cars per year. Eventually, their lack of mass production techniques caught up with them, and they went out of business in 1911. The next year, the factory was sold to a silo manufacturer. Anton Smith was so happy to be out of the auto business that in 1911, he made a bonfire using all the company records and files as fuel.

The **Smith Automobile Company** produced approximately one thousand vehicles, but very few remain. Terry Stafford, who had designed a few working autos as early as 1898, went on from the Smiths to form the **Stafford Motor Company.** They went on to produce autos from 1908 to 1915.

One customer of the *Stafford* was Harry S. Truman, who purchased one and mentioned it many times in his autobiography.

1911 Stafford

CHAPTER 17:

KENTUCKY'S DIXIE FLYER

THE KENTUCKY WAGON MANUFACTURING COMPANY was organized in 1878 to make buggies and carriages. Located in Louisville at 2601 South 3rd St., they became one of the largest carriage companies in North America. They produced popular wagons called the Kentucky, Old Hickory, and the Tennessee.

In 1914, they contracted with the **Hercules Motor Car Company** to produce auto bodies for a car they would produce. When the **Hercules Motor Car Company** went out of business in 1915, the **Kentucky Wagon Manufacturing Company** acquired the assets. From 1916 to 1923, the company designed and assembled the *Dixie Flyer* as well as a two-seater, the *Dixie Firefly*. The cars rotated between using Lycoming and Herschell-Spillman four-cylinder engines.

1922 Dixie Flyer

In 1916, the company name was changed to the **Dixie Motor Car Company.** They also produced a truck named *Old Hickory* and an electric truck, the *Urban.* In 1923, the company combined with the **Jackson Automobile Company** of Jackson, Michigan, and the **National Motor Car & Vehicle Corporation.** The last *Dixie Flyer* was released as the *National Model 6-31* in 1923. The company was eventually purchased, and the name changed to what it is today, the **Kentucky Trailer Company.** They presently make truck trailers.

1922 Dixie Flyer Touring Car Model H-S 70

Chapter 18:

The Louisiana Car Without a Chassis

In 1910, Charles Levy of New Orleans announced his plans to manufacture a car he called the *Simplicia*. It would be the "car without a chassis," an idea Levy picked up in France. Raising $2 million for the **Simplicia Motor Company**, the town of Adams, Massachusetts, expressed interest in having the factory built there. Reportedly, they managed to produce a few cars before disappearing altogether. This was the only recorded instance of a car company in Louisiana.

CHAPTER 19:

FROM MAKING WINDOW SCREENS TO AUTOS IN MAINE

EARLY IN MAINE'S HISTORY, THE auto named the *Whitney* was built in Brunswick, Maine, in 1899. The creator of the auto was R. S. Whitney, who was a jeweler and sold guns, ammunition, and sporting goods. He also ran a small machine shop and started working on a two-passenger runabout steam vehicle in 1898.

With the **Whitney Machine Company** in Brunswick, he built a second steam car in 1900, a four-passenger touring car, that he completed and sold in 1902. The third steam car, begun in 1903 and completed and sold in 1905, had a front-mounted boiler and an *all-steel* chassis, unusual for 1905. Whitney later received several patents for auto repair equipment.

The *Burrowes* was manufactured by the **E.T. Burrowes Company** of Portland, Maine, in 1905. At first manufacturing house screens, they decided to branch out into automobiles.

The first vehicle had a single chain-drive, and the engine was under the seat. With wire wheels and elliptical springs, the vehicle was well regarded, but not many were produced—maybe as few as twelve.

In 1908, the *Model E* came out and had a four-cylinder, L-Head engine with 30 horsepower. It was a better car but still not impressive enough to save the company. Not to worry; Edward Thomas Burrowes kept the screen business, and by the time he retired in 1912, it was the largest screen manufacturer in the world. He was also doing a good business with billiard tables and folding card tables.

1908 Burrowes Model E; Courtesy of Maine Transportation Museum

Chapter 20:

The Burns, Crawford, and Moller Car Companies of Maryland

Brothers W., R. H., and A. G. Burns produced a high-wheeler auto from 1908 to 1912 in Havre de Grace, Maryland. They added a double-chain drive, a two-cylinder, air-cooled engine, and friction transmission to the carriages they had been manufacturing as the **Burns Brothers Company.** By 1912, the vehicle still seemed unacceptable to the public, even after revamping it by using standard wheels, bringing their buggies closer to the ground. No matter what they did, their "motorized carriage" was still unsuccessful, and they went out of business.

The first *Crawford* was a two-cylinder, 10-horsepower, water-cooled gasoline motor car manufactured by the **Crawford Motor Company** of Hagerstown, Maryland, from 1904 to 1923.

The company founders were Mathias Moller, who was a pipe organ manufacturer; Robert Crawford, who was a bicycle builder and developed the car's engine; and Hagerstown businessman Henry Holzapfel. It was originally named the **Crawford Automobile Company.**

1909 Crawford Model H Light Touring Car

Car production included building two small runabouts in 1904 and five in 1905. In 1906, they added a six-cylinder engine to the vehicle, and production totaled forty-one vehicles. The early cars used Rutenber engines, and later models had Continental engines. The cars switched from planetary transmissions to a sliding gear in 1908.

Annual production peaked at 275 cars in 1910. From 1916 on, all *Crawford*s were produced with six-cylinder engines. By late 1916, production had dwindled to 104 cars. In 1917, only thirty-eight were built because of the first year of World War I shortages. A twelve-cylinder *Crawford* truck was promised, but it never materialized. In 1922, fifty-four *Crawford* cars were built, and in 1923, the last cars were assembled with the remaining parts. The end of the *Crawford* auto was announced in March 1924, with production to change to the *Dagmar* auto.

Named after Mathias Moller's daughter, the *Dagmar* auto began production in the summer of 1922. It was at first meant as a sporty companion car to the *Crawford*. It had a six-cylinder, 70-horsepower Continental engine with a 138-inch wheelbase. In 1924, the name of the company was changed to the **M.P. Moller Motor Car Company.**

In 1925, a smaller version of the *Dagmar* was released. It had a 120-inch wheelbase and was going to be called the *Petite*, but it came to be known as the *Baby Dagmar*. The total car production for 1923 was 135, and in 1924, a new *Dagmar* was presented to the winner of the Miss America beauty pageant.

Both the **Packard Motor Company** and the Danish Royal House expressed complaints to the **M.P. Moller Motor Car Company.** Packard objected to the *Dagmar's* copying of its radiator and hubcaps, and the Danish Royal House, through the Danish Embassy, objected to their coat of arms being used as the *Dagmar's* emblem.

Some of the mottos used to advertise the *Dagmar* included "An Automobile Classic," "The Ultra Sport Car," and "The Car of Swagger Individuality." Production of the *Dagmar* stopped in 1927. The last one built was a seven-passenger limo, custom-built for Moller to take to Europe when he visited his native Denmark.

In 1924, they received a contract to produce taxicabs. In 1927, the *Aristocrat* was built by the **Moller** organization in Hagerstown. It had a Buda-designed engine and was built as a taxi, although it differed in its radiator from other **Moller**-built cars. Other taxicab models included *Astor, Blue Light, Five-Boro, Luxor, Paramount, Super Paramount,* and *Twentieth-Century.* Production continued until 1931.

Three separate companies had the name Maryland in their company's name. The earliest was the *Maryland Steamer* in 1900 and 1901. The *Maryland Steamer* was a two-cylinder auto with a single-chain drive made by the **Maryland Automobile & Manufacturing Company** in Luke, Maryland, and sold between $900 to $2,500. In 1901, they requested to go into receivership.

In 1907, the *Maryland* was built by the **Sinclair-Scott Company** in Baltimore. This company had previously produced the Ariel bike, which used many of the same parts. In four years, until 1911, 871 cars were built until the parts backlog was exhausted. By

1912, the company returned to its original occupation of manufacturing food-processing machinery.

In 1907, the **Carter Motor Car Corporation** was formed in Hyattsville, Maryland, to manufacture the *Carter Two-Engine Car*, named after company founder Howard Carter. As a fail-safe measure, the car was equipped with *two* engines in case one stopped working. By 1908, it was obvious from the vehicle's poor sales that Americans preferred to take their chances with one engine. The manufacturers of the *Washington* auto moved into the factory in 1909.

CHAPTER 21:

THE GAS, ELECTRIC, AND STEAM CARS OF MASSACHUSETTS

SYLVESTER ROPER BUILT MARINE STEAM engines beginning in 1859. *Scientific American* did an article on his engine experimentation and a two-passenger, four-wheel vehicle with a 2-horsepower steam engine, capable of 25 mile-per-hour speeds, that he built. The steam engine was powered by coal, carried under the seat.

Roper built at least ten steam vehicles, one of which was exhibited to the W.W. Austen Carnival as the "greatest wonder of the world." In 1896, at the age of seventy-three, he crashed while driving at the Charles River Track and died shortly thereafter.

———

Frank and Charles Duryea were pioneers of the auto industry, responsible for building some of the United States' first autos. They were born in Illinois and moved to Washington, D.C., where they started a bicycle business. Moving again to Springfield, Massachusetts, the brothers used the Springfield Library to read about and research the internal combustion engine.

By 1893, the Duryea brothers had developed the first working American, gasoline-powered automobile. It was built from a wagon that the brothers powered with a single-cylinder, 4-horsepower engine. They called it a motor wagon. Charles was the chief engineer, while Frank built and tested the auto in 1892 and 1893.

On September 20, 1893, the brothers successfully road-tested their auto. On November 20, 1895, Frank drove the car and won the first American car race, beating three Benz cars from Europe.

The race helped sales of the brothers' **Duryea Motor Wagon Company** autos as they became the first to commercially sell autos in America. The brothers split up, and Frank went on to become part owner of the **Stevens-Duryea Company.**

The 1909 Stevens-Duryea Seven-Passenger Touring Car

1903 Stevens-Duryea Roadster

Early Duryea vehicle and Frank Duryea

Charles Duryea, old and young

John Northway built engines in Detroit and sold his concern to **General Motors**. He moved to Canton, Ohio, and produced the *Crescent* auto.

Coming to Natick, Massachusetts, in 1918, he got investors to back him in starting the **Northway Motor Company** to manufacture cars and trucks. The first vehicle manufactured was a truck, and in 1921, the *Northway*, a six-cylinder, 61-horsepower vehicle, was available in touring, sedan, sport, and coupe styles. In 1922, the company's last year, the car was essentially the same. John Northway went on to manufacture *Maxim* fire engines.

Plant of the Northway Motor & Mfg. Co., Detroit, Mich.

In 1893, John Piper and George Tinker started the **Waltham Manufacturing Company** in Waltham, Massachusetts, to build bicycles. Their two-cylinder auto had a single-chain drive and used a tiller for steering. Their cars, built until 1900, were the *Waltham* and the *Orient Buckboard*.

The **Waltham Manufacturing Company** was absorbed by the **Metz Company** and produced a car named the *Metz* until 1922. It was originally sold by the Metz Plan, in which the consumer would receive fourteen packages in the mail to assemble the *Metz* at home. Later models were sold in the usual manner.

The Waltham

The 1910 Metz of Waltham, Massachusetts

In 1916, the *Metz* was retooled into the *Master Six* with a 45-horsepower, six-cylinder engine. In 1922, the car was renamed the *Waltham,* bringing the company full circle.

In 1897, the same year the Duryeas were debuting their gasoline vehicle, the Stanley twins, Freelan and Francis, started manufacturing their *Stanley Steamer* in Newton, Massachusetts. It was a successful auto, even though at first it took twenty minutes for the boiler to heat up enough to start the car. This was an improvement over many steam cars that took up to an hour to be hot enough. The *Stanley Steamer* was manufactured until 1912 and was the last of the great steam cars.

The Stanleys started selling their steam-powered auto to the public in 1899. Steam power was supplanted by gasoline engines but not before the Stanleys warned consumers that when they bought a car powered by gasoline, they were using an "internal explosion engine."

The 1902 Stanley Steamer and the 1911 Stanley Steamer

1911 10 h.p. STANLEY STEAM CAR

The Stanley twins had used two different types of steam loco-motion in their autos during their tenure at the **Stanley Motor Carriage Company.** The original design was sold to the **Locomobile Motor Company of America.**

1905 Stanley Steamer, 1912 Stanley Steamer

1908 Stanley Steamer Runabout

While many auto pioneers were brothers, the Stanleys were identical twins. When one twin would be in the act of getting a ticket, the other Stanley twin would drive by in an identical vehicle, confounding the police to no end! In 1906, the world speed record, 127.7 miles per hour, was set in a *Stanley Steamer.*

The twin brothers were very close, and in 1918, when Francis died in a car accident, Freelan sold the company. Francis was killed after swerving into a woodpile to avoid hitting some people on farm vehicles.

The **Knox Automobile Company** had a respectable showing in the annals of car companies, manufacturing autos from 1900 to 1914. They built tractors and trucks until 1924. They were founded in Springfield, Massachusetts, in 1900 by Harry Knox.

The first *Knox* vehicle was a three-wheeler with a 4-horsepower, single-cylinder, air-cooled engine. The first four-wheeler was in 1901. Single-chain drive was replaced by double-chain drive in 1905. A water-cooled engine became an option in 1908. In later years, large four- and six-cylinder cars were available.

Harry Knox went from the **Knox Automobile Company** to forming his own company in 1905 across the street from his old one in Springfield, Massachusetts. He called his new company the **Knox Motor Car Company**, which very much displeased the owners of his previous company. In 1907, their court action forced Harry Knox to rename his company the **Atlas Motor Vehicle Company.**

The first vehicle Knox manufactured in his new company was the *Atlas,* a two-ton truck. In 1907, he struck a deal with the San Francisco, California, manufacturers of a vehicle called the *Sunset.* They needed money to rebuild their factory after the 1906 earthquake. Knox licensed the two-stroke vehicle and produced his own version in 1907, a runabout also named the *Atlas.* They used the motto "The Perfected Two-Cycle Engine." In 1908, he marketed a delivery van and a taxicab, both using the two-cylinder, two-stroke engine, and also began offering three- and four-cylinder engines.

In 1912, the Knight engine was adopted, and the name of the models changed to *Atlas-Knight.* In 1913, the company went into bankruptcy. Afterward, Walter Knox joined the Lyons brothers in building the *Lyons-Knight* in Indianapolis, Indiana.

In 1906, the **Climax Electric Works** company began in the henhouse that belonged to Levi W. Flagg in New Salem, Massachusetts. William E. Taft had already promoted a steam car named after him and decided to ally with Flagg to build an electric car. When the electric cars built became too costly for the average consumer, they switched to gasoline vehicle research.

The car that hit the market was a two-cylinder, 18-horsepower, gasoline-powered auto with a ninety-inch wheelbase. By 1908, the company sold enough cars to move out of the henhouse. The firm was in business until they went bankrupt in 1911.

The *Crestmobile* was built in Cambridge and Dorchester, Massachusetts, from 1901 to 1905 by the **Crest Manufacturing Company.** Their first offering was initially a three-wheeled vehicle with a single-cylinder, chain-drive, two-seater auto with air-cooled engines. In 1903, a shaft drive replaced the chain drive.

Although they tried promoting that "one cylinder is more efficient than two," the advertisement wasn't working, and the company added two-cylinder vehicles in 1904 and 1905.

In 1905, the **Crest Manufacturing Company** was absorbed into the **Alden-Sampson Manufacturing Company** of Pittsfield, Massachusetts, a commercial vehicle manufacturer. They continued to sell leftover *Crestmobiles* through 1907. The **Alden-Sampson Manufacturing Company** joined Benjamin Briscoe's **United States Motor Company** in 1910. They didn't make the cut when the consortium stopped manufacturing the commercial vehicles to concentrate on the *Maxwell*, which became the *Chrysler*.

The Grout brothers Carl, Fred, and C. B. were the sons of William Grout, who had founded the White Sewing Machine Company (along with Thomas H. White) in the late 1850s. Since they weren't interested in sewing machines, William Grout set up his sons as the **Grout Brothers Automobile Company**, just as Thomas White set up his sons in the auto manufacturing business. The White sons went on to make successful steam cars. The Grout brothers weren't entirely truthful in their publicity brochures. They stated that in

1896, they built the first automobile factory in the United States, which was far from true.

The first steam car of the Grouts made its exit from the factory in Orange, Massachusetts, in 1901 and was named *New Home*. By 1904, business was good, and eighteen steam cars were being produced weekly. Some were shipped overseas and marketed under the name of *Weston*.

Gasoline vehicles were introduced by the Grout brothers in 1904 and by 1906 were exclusively produced, with the steam vehicles being phased out.

1904 Steam-Powered Grout

In 1907, the Grout brothers' father, William Grout, took over the company, stating that he didn't like the way it was being run. The sons tried to have their father declared "senile" (he was seventy-four), but they were unsuccessful, and the sons left the state. The company was left in chaos when William died on April 20, 1908.

Three Orange, Massachusetts, businessmen purchased and reorganized the company, only to have it retreat back into chaos when two of the three businessmen died. Production of the autos ceased in 1912. The plant was sold at auction in September 1913. In August 1914, it was announced that the factory had been leased by the **Red Arrow Automobile Company**.

The 1900 *Porter Stanhope* was described in their advertising as "The Only Perfect Automobile." Produced by the **Porter Motor Company** with the factory in Allston, Massachusetts, and the headquarters at 950 Tremont Building, Boston, Massachusetts, the *Porter* was invented by Major D. Porter. It had a two-cylinder, single-acting steam engine that could use either kerosene or gasoline for fuel.

The *Porter Stanhope* was a two-passenger vehicle with a runabout style and *all-aluminum* body. It was the first car with an all-aluminum body and weighed only 550 pounds. The *Porter Stanhope*, sometimes called the *Portermobile*, had a safety device— if you took your hand off the controlling lever, the car shut off. The car was manufactured in 1900 and 1901.

Chapter 22:

The Multitude of Car Companies of Motown, Michigan

Michigan's Motor City, Detroit, had over one hundred car companies within its borders, but what separates Michigan from all the rest is the number of cities in Michigan, besides Detroit, that produced autos.

Map showing proliferation of car factories in the Detroit area

The second-largest city in Michigan, Flint, was the birthplace of **General Motors** and several other auto companies. Lansing, Jackson, Kalamazoo, Port Huron, Grand Rapids, Alpena, and Saginaw all had car companies.

Often, three reasons are given to explain why Detroit became the place where American car production centered. The first reason is that all the car companies were near each other, allowing more collusion on new ideas between Henry Ford, who had the idea of cars for the masses, and Ransom Olds, who proved that such a car could be built.

Ransom E. Olds was the first person to drive a car down the streets of Lansing. He experimented with steam and electric cars before settling on gasoline engines, which he sold separately before trying to manufacture an auto. When he was ready to build his auto, Olds chose Detroit to build it in since he heard that lots of engineers and mechanics lived in the larger city. Moving there, he built a factory for the assembling of autos, said to be the first. He also suffered one of the first auto company fires. All his prototypes burned up in the fire except for one, driven out of the burning factory in the nick of time by James J. Brady. (Brady also coined the *Oldsmobile* name and was later one of the founders of the **Hudson Motor Company**.) That car was the *Curved Dash Olds*, which the **Olds Motor Works** built to great success, becoming the number-one-selling automobile in the early 1900s.

The only thing was, the car was built in Lansing, not Detroit. When the Lansing Chamber of Commerce heard about Olds's factory misfortune, they offered him very favorable terms to return to Lansing and build his autos there.

In 1904 and 1905, the company had the *Oldsmobile Model N "French Front"* touring runabout, which cost $100 more than the *Curved Dash* at $750.

Since the *Curved Dash* car was so successful, Ransom Olds couldn't understand why his stockholders wanted to stop building the *Curved Dash Olds* and instead switch to building luxury cars, like most auto companies of the early 1900s.

1904 Oldsmobile Model N

When the stockholders disagreed, Olds left the company he started and started a new one with the name **REO Motor Company**, named after his initials, Ransom E. Olds. His company was very successful building trucks as well as cars. Their most famous was the *REO Speedwagon*. It wasn't long before **REO** sales surpassed those of the **Olds Motor Company** and **Olds** stock was selling at such a low price that Billy Durant was able to buy a controlling interest in it and add *Oldsmobile* to his stable of cars at **General Motors**.

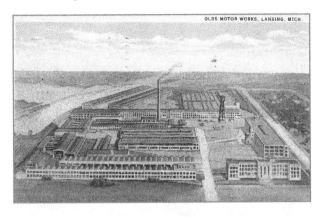

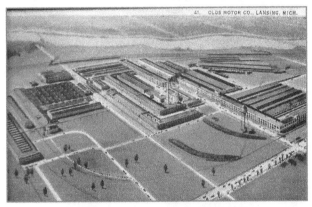

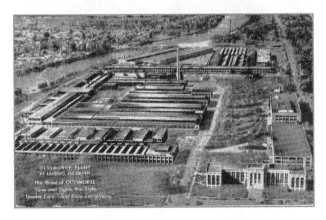

*Views of the Lansing factory of the **Olds Motor Works**, the **REO** factory, and the REO Speedwagon*

Henry Ford hated the delivery vehicles his first company, the **Detroit Automobile Company**, manufactured. For one thing, they wouldn't start! All agreed to disband the company, and many of the same investors signed on for Ford's second company, the **Henry Ford Company**, but Henry Ford and the board of this company didn't see eye to eye and agreed to part company.

Since the investors wanted to get some of their money back, they asked Henry Leland, proprietor of engine and machine builders **Leland & Falconer**, to appraise the equipment. Instead of appraising the equipment, Leland talked the investors into building HIS car instead since they had all the infrastructure needed to construct cars already. The investors agreed and decided the new company and car should be named after the founder of Detroit, Antoine de la Cadillac, so the **Cadillac Motor Company**'s *Cadillac* was a success from the start, a lot due to the dealership network set up by William Metzger, and *Cadillac* autos instantly became top sellers.

The 1903 and 1912 Cadillacs

*Leland & Falconer Manufacturing on Milwaukee Street
in the Milwaukee Junction area of Detroit*

The **Cadillac Motor Company** was absorbed by Billy Durant's **General Motors** and became the luxury marque of **GM**. After **Cadillac** founder Henry Leland worked on Liberty aircraft engines during World War I, he used his new expertise to develop the Lincoln engine and form the **Lincoln Motor Company** to make *Lincoln* autos. Just as Leland in a way took over Ford's second motor company, Ford bought Leland's second company, and **Lincoln Motor Company** became the luxury marque for the **Ford Motor Company**.

1948 and 1929 Lincolns

Lincoln Motor Factory in Detroit

Lincoln Motor Company factory

Byron Carter invented a gasoline vehicle in the late 1890s but switched to steam-powered vehicles in 1900. After operating a company selling steam vehicles, in 1903, Byron J. Carter put together financing for the **Jackson Automobile Company** to build steam and gasoline autos. The steam car was called the *Jaxon* and was produced in 1903 only. They continued with gasoline cars. Byron Carter was upset when the company wouldn't accept his friction-drive transmission, and he left to form his own company and build the *Cartercar* (with friction-drive transmission).

Ad for Byron Carter's Jaxon

*The Cartercar of Byron Carter, who later died of gangrene of the jaw after
a stranger's car crank kicked back and hit him in the jaw.*
See page 176 for more on Byron Carter.

The **Jackson Automobile Company** persevered without its
founder until 1924. The car's motto was "No Hill Too Steep, No
Sand Too Deep." The original car in 1903 was a one-cylinder runabout; in 1904, it doubled to a two-cylinder; in 1906, they brought
out a four-cylinder; in 1913, a six-cylinder; and finally, in 1916, a
car with a V-8 engine made its debut.

From 1909 to 1911, the company produced a high-wheeler car
called the *Fuller.* They also manufactured a "companion car" to the
Jaxon called the *Orlo* in 1904.

Maybe one of the most ridiculed cars ever released was the *Duck,* also called the *Jackson Back-Seat Steer,* which was a car steered from the back seat. In 1917 and 1918, many of the *Jaxons* were marketed as *Wolverines.* As the factory moved over to defense work and manufactured fewer vehicles, many of their dealers began to sell *Jordans.* In 1919, the factory concentrated on war production.

In 1920, the **Jackson Automobile Company** struggled as it returned to peacetime auto production. The 1920 autos had production flaws, and the company was having trouble getting credit. By 1922, the company was releasing the last of the *Jaxon* marque autos. In 1923, the company merged with **National** of Indiana and **Dixie Flyer** of Kentucky to form the **Associated Motor Company**. This new company was gone by 1925.

Factory and headquarters of the Jackson Automobile Company in Jackson, Michigan

1908 Jackson auto

The *Marion-Handley* was built by the **Mutual Motors Company** of Jackson, Michigan, from 1916 to 1919. Two models were offered: a four-seater roadster and a touring car. These were powered by a Continental six-cylinder engine. Wire wheels were an option, and the auto was very popular its first year.

1917 Marion-Handley Roadster

The **Marion Motor Company** of Indianapolis had been a builder of race cars and the *Bobcat Roadster*. Perpetually underfinanced, the Indianapolis company started in 1904 and in 1915 was purchased by J. J. Handley; the name of the cars changed to the *Marion-Handley*.

Handley also bought the **Imperial Automobile Company** of Jackson, keeping the two companies separate. **Imperial** stopped producing autos in 1915. The *Marion-Handley* went on until 1918.

———

Brothers George N. and T. A. Campbell started the **Imperial Automobile Company** in Jackson, Michigan, in 1908 when they added motor cars to their already prosperous carriage business. They manufactured mid-sized, four-cylinder cars with coachwork by the **Beaudette Company**, who also built bodies for **Buick** and **Ford.**

Imperial Motor Company logo

The 1910 *Imperial Model 45* was a four-cylinder car with 50 horsepower. Its wheelbase was 117 inches, and it sold for $2,000. Further models had varying styles and cylinders.

In 1912, the **Imperial** factory burned down, but the company quickly acquired an unused *Buick* truck factory in Jackson. In 1917, the Campbell brothers sold their operation to the **Marion-Handley Company**, and the *Imperial* marque was discontinued.

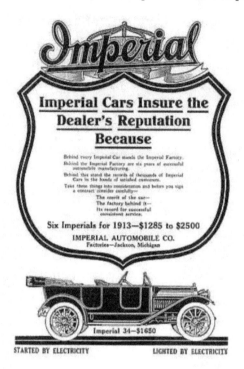

Imperial Automobile Company advertisement

1929 Oakland Model 212 All American Landaulette Sedan

Alanson Brush took advantage of being Henry Ford's neighbor and commiserated with him on autos. Brush went on to start the **Oakland Motor Company** and the **Brush Motor Company.** The **Oakland Motor Company** started as the **Pontiac Buggy Company** and started building autos in 1907. Alanson Brush's partner in this venture was Edward Murphy, and their first auto contained two cylinders. The next year, that was doubled to a four-cylinder and began a tradition of an improved auto each year.

In 1910, the marque was purchased by Billy Durant for his fledgling **General Motors**. One of the models, the *Pontiac* (named after the city and the Indian chief), became the new name of the company, and the **Pontiac Motor Company** became a long-lasting division of General Motors, lasting until 2010.

1914 Oakland

The Liberty Motor Car Company

The **Liberty Motor Car Company** began in Detroit in 1916 with an initial capitalization of $400,000. With a six-cylinder engine and a range of models available, including roadster, touring, brougham, town car, and coupe, in the first year, they manufactured 733 cars.

Liberty Motor Car Company (Independence Hall copy to the left in the picture)

The next year, they doubled production and stayed competitive with their *Liberty* cars. In 1921, they produced eleven thousand cars. When the company moved to a new plant in 1922 (with the outdoor façade a copy of Philadelphia's Independence Hall), they started developing financial problems and by 1923 were acquired by the **Columbia Motor Car Company**. They put out the same *Liberty Six* car as the year before, but by 1925, the new company was bankrupt. The factory was sold to the **Budd Manufacturing Company**.

The **Columbia Motor Car Company** was incorporated in 1916 with **King Motor Company**'s J. G. Bayerline and William Metzger as the VP. They produced a popular car, the *Columbia Six*, that was the first to feature a radiator temperature control and other high-quality parts.

Producing over three thousand cars per year by 1923, the company was doing great until it got overenthusiastic and bought the **Liberty Motor Company**, taking over their plant. It was too much too soon, and the company went bankrupt the following year.

The Liberty Motor Company headquarters, a copy of Independence Hall

One of the most venerable of the Detroit auto companies, from 1908 to 1940, was the **Hupp Motor Car Company**. Started in 1909 by Robert "Bobby" Hupp and four investors, they started in a small building at 345 (now 1051) Bellevue Avenue in Detroit.

Robert Hupp was from Grand Rapids and moved to Detroit to work for **Olds Motor Company**. He then worked for **Ford Motor Company**, teaming with John Dodge and C. Harold Wills on the *Ford Model K*. He quit in 1907 to work for the **Regal Motor Company**, and then, in 1908, he developed his four-cylinder, lightweight auto, the *Hupp 20*. By 1909, the car was in production, and at the end of the year, his brother Louis joined him in his auto venture.

Hupp Motor Car Company

Hupp Motor Company factory

The *Hupp 20* was very popular and sold a phenomenal five thousand autos in 1910. It was even more popular when three men drove it around the world. The car was like an all-purpose terrain vehicle as it traveled across sand in China and roads that were little more than cattle trails. Before the car got back to the United

States, the orders started pouring in, but the company was no longer making the small, lightweight vehicle. Instead, the board had decided that the company should manufacture a more luxurious vehicle (much like the *Oldsmobile* board) and retooled the factory to do so.

1910 Hupp-Yeats Electric Car

Robert Hupp quit in a huff, and his brother Louis followed. The two had started the **Hupp-Yeats Motor Company** with an investor (Yeats) and manufactured a successful electric car in 1911 before Hupp quit the **Hupp Motor Company**. The original **Hupp Motor Company** sued to get Robert Hupp's name off the car and company per their prior agreement that Hupp wouldn't name any more car companies after himself. Robert Hupp informed them that the company was named after his brother Louis.

The controversy continued until the Hupp brothers changed their company name to the **RCH Corporation** (Robert Craig Hupp's initials). This was much like Ransom E. Olds having to use his initials for the **REO Motor Company** since the **Olds Motor Company** wouldn't let him use "Olds" in his new company name.

Hupp had put all his companies into one big one, the **Hupp**

Corporation, to hold the many parts factories that Robert Hupp had purchased control of. This company name was changed to the **RCH Corporation** upon losing the Hupp name lawsuit. Robert Hupp bought out his first *RCH* car in 1911. The company went under in 1914, and Robert Hupp joined the **Monarch Motor Company** until 1917. After that, the Hupps spent their time running their auto parts empire.

1912 RCH Touring Car

Meanwhile, the original **Hupp Motor Car Company** moved to larger quarters at Jefferson and Concord streets in Detroit. They sold this factory to the **King Motor Company** to move to still larger quarters at Mt. Elliott and Milwaukee streets in the Milwaukee Junction area of Detroit. In 1911, they pioneered the use of steel in auto bodies with their *Hupp 32* model, said to be the first American, all-steel car.

1911 Hupmobile

The *Hupmobile* provided strong competition to *Chevrolet* and *Ford* in the 1920s. In 1925, they brought out an eight-cylinder luxury model. By 1928, the annual **Hupp** production was over sixty-five thousand autos. The company purchased the **Chandler Motor Company** of Cleveland, Ohio, mainly so they could use the factory facilities of the defunct company.

1914 Hupp Runabout

Hupp execs pose by factory

When the Great Depression happened in the 1930s and killed off many car companies, **Hupp Motors** stuck in there, still selling 1,600 cars per year after former years of selling six thousand and seven thousand cars per year. They developed new styles and survived despite lower sales. They struck a deal with **Graham-Paige** to coproduce a car called the *Skylark* but were unable to make more than three hundred. Finally, by 1939, a combination of delinquent Detroit city taxes and fighting a takeover by the undesirable stock manipulator Archie Andrews caused the company to have to stop manufacturing autos. Instead, they switched to war production for World War II and after the war made auto supplies but not complete autos. Over half a million cars were manufactured during the lifetime of the marque. They finally morphed into **Hupp, Inc.** and now make industrial heaters and air conditioner units.

1942 Skylark

1934 Hupmobile (Courtesy of Stahl's Automotive Museum)

The **Poss Motor Company** in Detroit was incorporated in the summer of 1911 with $250,000 to manufacture autos. They reportedly produced a few commercial vehicles in 1912 but no passenger cars.

Poss Motor Company factory

The **Monarch Motor Company** started in Detroit in 1913 when Joseph Bloom started using the **Carthartt Motor Car Company** factory that had recently been vacated. His brother-in-law, Robert Hupp, of *Hupmobile* fame, had designed the four-cylinder cars, called *Monarchs*, available in touring and runabout styles. Expected money didn't come through, though, and even though the company released an eight-cylinder car in 1916, they declared bankruptcy later that same year.

In 1907, Martin Kessler built an auto engine for the **Chalmers Motor Company.** After working for a few other auto concerns, he started his own company, **Kessler Motor Company**, in 1917, and did well manufacturing aircraft engines during World War I.

The Kessler Motor Company factory

After the war, in 1920, the company announced their first passenger car, a 70-horsepower, four-cylinder touring car, for $1,995. It was called the *Kessler Super-Charge Four.*

In 1921, they brought out a new car with a new company—**Kess-Line Motors.** This supercharged auto was called the *Kess-Line Eight* and had an eight-cylinder engine capable of generating 100 horsepower. The offspring company moved into the factory of the old **Liberty Motor Company.** The company didn't manufacture many autos, although they were in business until 1927, when they went into receivership.

The Kess-Line factory in Detroit

Regal Motor Car Company built a five-passenger car called the *Regal* in Detroit from 1908 to 1918. For the first years using smaller, four-cylinder, 25- and 30-horsepower engines, they switched to manufacturing their own 25-, 35-, and 45-horsepower engines for their cars from 1910 to 1914.

Regals had water-cooled engines and sliding-gear and shaft drives, selling for $1,250. They began using an "underslung" design (frame below the axle), and overall production climbed to 7,500 vehicles manufactured by 1913 at the rate of about one thousand per year. They also exported autos to England. Models included the *Colonial Coupe,* which sold for $1,250; the touring car for $950; and the roadster for $900.

The **Regal Motor Car Company** went into receivership in 1918 due to shortages caused by World War I.

*The Regal factory at 201 Piquette Street,
bordered by Hastings Street*

Regal factory

A 1912 Regal Model N 25-horsepower roadster

1910 Regal stuck in the road

The *Rockne* was an auto manufactured by the **Rockne Motors Corporation** from 1931 to 1933. Controlled by **Studebaker** and built in Detroit, the car was named after the famous Notre Dame football coach.

Plans were for Knute Rockne to do promotion for the car, but he died in a plane crash right before the car's debut in 1932. The cars had six-cylinder engines and a wide range of body styles, including roadster, coupe, and sedan. A notable feature was the mud flap. It sold for $585 to $675 but was discontinued in 1933 after 30,293 were sold. The factory was dismantled, and everything that could be used was shipped to the **Studebaker** factory in South Bend, Indiana.

Charles Brady King, the inventor of the pneumatic jackhammer, drove his first car, one he constructed himself, down the streets of Detroit in 1896, beating Henry Ford by months.

King was born in California and graduated from Cornell with a degree in engineering. He moved to Detroit, eager to work on automobiles, and worked at **Oldsmobile** in 1900 until their factory burned down. He organized and became chief engineer of the **Northern Motor Company** from 1902 to 1908 and designed the famous *Silent Northern* car while there.

King went to Europe to study auto engineering from 1909 to 1910. When he returned, he started the **King Motor Company**, releasing the first car to have left-hand steering. He is also considered to be the first to put together a viable V-8 engine. He retired from **King Motor Company** in 1912, although the company continued, producing over one thousand models. He was a wealthy man, having a large income from the sixty-four machine patents he owned, including forty for automobiles.

Charles B. King

Northern trading card, the King factory

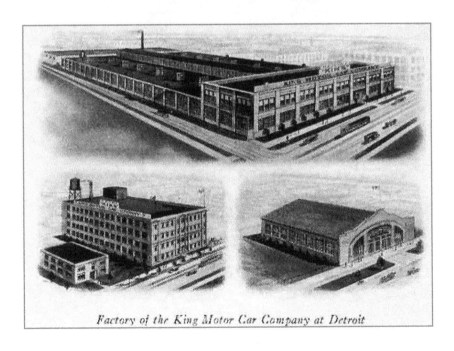

Factory of the King Motor Car Company at Detroit

The **Northern Motor Car Company** was one of the first in Detroit, founded by Jonathan Maxwell, Charles B. King, and Charles Palms, financier and grandson of the onetime richest man in Michigan (before Henry Ford), Francis Palms. Their first car, the two-cylinder, two-seater runabout, the *Silent Northern,* was their most popular model. It was similar mechanically to the curved-dash *Oldsmobile* but with a straight dash. In 1906, King developed a four-cylinder, 18-horsepower vehicle with air-operated clutch and brakes. When King resigned in 1909, the company merged with **E-M-F.**

1917 King

1912 Ann Arbor Touring Car

In 1911, the **Huron River Manufacturing Company** of Ann Arbor, Michigan, manufactured a two-cylinder convertible touring car that had an easily removable back seat if the space was needed for packages or luggage. The price for two seats and sideboards was $950. You could get another seat for an additional $25 and a canopy top with a side-curtain storm front for another $50.

It had a top speed of twenty-five miles per hour—maybe that's why it was only produced one more year, until 1912. In 1913, the **Star Motor Company** moved into the Ann Arbor factory to produce commercial vehicles.

Charles W. Nash was one of the best examples of a poor boy working his way up in the world in a Horatio Alger type of way. He worked as an indentured servant until thriving in his own businesses. He eventually became head of **General Motors** in Flint and then personally bought the **Thomas B. Jeffries Company** (makers of the *Rambler)* and renamed it **Nash Motors Company**.

Nash Metropolitan

Building their cars in Kenosha, Wisconsin, site of the *Rambler* factories, **Nash** autos continued from 1918 to 1942, including many famous marques centering on the six-cylinder car like the *Super Six* and the *Special Six*. Other famous **Nash** brands include the *Ambassador* and the *Metropolitan*.

Nash trading card

C. Nash Flint birthplace

The Nash Motors Company factory in Kenosha, Wisconsin

The Nash plant, later AMC HQ

In 1902, C. H. Blomstrom built a gasoline car he called a *Blomstrom*. It was a small, runabout car manufactured by his Detroit-based **C.H. Blomstrom Motor Company**. He had earlier, in 1897 and 1899, built prototype vehicles. In 1904, he changed the name of the car to the *Queen*. The auto had one cylinder and could generate 8 horsepower. About 1,500 *Queens* were manufactured until 1906, when the company was combined with the **Car De-Luxe Company**. C. H. Blomstrom was also involved with the vehicles called the *Rex Cycle-car*, the *Frontmobile*, and the *Gyroscope*.

The *Austin* of Grand Rapids, Michigan, was first built in 1903 by the **Austin Automobile Company**. James Austin made his money in lumber, which enabled him to purchase **Michigan Iron Works** in 1900 for both investment purposes and to give his mechanically inclined son, Walter, a place to tinker. By 1903, Walter had produced thirteen *Austins*, a two-cylinder, 25-horsepower touring car with a ninety-inch wheelbase.

Each succeeding year brought a larger engine with more horsepower so that by 1907, four-cylinder cars with 60 horsepower and six-cylinder cars with 90 horsepower were available. In 1911, the car featured the steering wheel on the left and electric lights. In 1913, they had a large touring car with a 142-inch wheelbase, known as the *Highway King*. The 1917 version had a twelve-cylinder engine.

Although they tried marketing a smaller, more affordable car, they were known for the grander versions. Celebrity owners of *Austins* included boxer Jack Johnson and publisher William Randolph Hearst.

Austin automobile ad

1909 Austin

The company stopped producing cars in 1920 during the post-World War I recession. Austin and his father went into real estate speculation together, culminating the father-and-son business relationship, which lasted over fifty years. Walter patented a hydraulic transmission in 1950 and died in 1965, less than a year away from turning one hundred.

In 1909, the Muskegon, Michigan, Chamber of Commerce built a factory for the **Gary Motor Car Company**, which failed before the company could even take possession. In November 1909, the **Henry Motor Company** was happy to find a brand-new, unoccupied auto factory and moved in. David W. Henry was the designer and had previously worked for the **Columbia Company** in Hartford, Connecticut, for over ten years.

The *Henry* was a four-cylinder, 35-horsepower touring car. The 1911 and 1912 models included touring, tonneau, and torpedo models as well as a roadster named the *Roadstar* in a Muskegon newspaper contest.

In 1911, Michigan Lieutenant Governor John Q. Ross took over as president of the company. David Henry left the company after an argument and went to Iowa to help build the *Colby*. Lack of capital caused the company to go into receivership in 1912 after they had produced over six hundred cars. The company was sold to a Philadelphia firm for $9,000.

The *Herreshoff* auto was built by a company known for their marine engines and yachts, the **Herreshoff Motor Company**. Charles Herreshoff started his car division in Detroit in 1908. They operated out of the old **Thomas-Detroit Company** factory. The car manufactured in 1909 was a four-cylinder, 24-horsepower car, the *Herreshoff*, available in touring and roadster styles. By 1910, runabout styles were available as well as six-cylinder engines.

The 1913 Herreshoff Coupe

In 1911, the company broke ground for a new Detroit factory at Belmont and Trowbridge Streets, but the neighbors protested so vigorously that the factory wasn't built. The car company "had problems with its Lycoming engines," said Charles Herreshoff as he quit the company in March 1914. Faced with financial difficulties, the company was sold later that year. Charles Herreshoff later built a car called the *Harvard* in Troy, New York.

William Anderson sold his successful carriage company in Port Huron and moved to Detroit in 1907 to build cars. He ended up with the most successful electric car company in the United States, the **Detroit Electric Car Company**, by the time of their demise in 1939; it had a battery that would allow the auto to go 211.3 miles without a recharge. When the car did need to be charged, they had recharge stations all over Detroit—one still survives near the Belle Isle Bridge.

Detroit Electric Car Company recharge stations

When they ended official production in 1939, the parts were all sold to a dealer who continued to assemble Detroit Electrics by special order. Called the "smallest auto factory in the world," it was located on the second story of a building on 10th Street in Detroit.

10th Street factory (second story) for Detroit Electric, 1940

The **Dragon Automobile Company** had top engineers design their car. It had four cylinders and cost only $2,000, which in 1909 was a great deal of money. Unfortunately, it often had mechanical problems. Even more unfortunate was the fact that the company had borrowed $136,000 from a Philadelphia bank using two hundred *Dragon* autos as collateral. The only thing was, they didn't have two hundred autos! It's known that *Dragon* autos were produced, but the exact number is uncertain. By 1908, the **Dragon Automobile Company** were bankrupt and had disappeared.

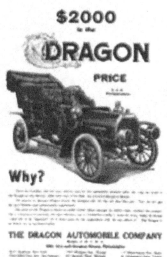

Dragon Auto Company logos, car, and ad

The **Paige-Detroit Motor Car Company** started with Frederick Osgood Paige, former president of the **Reliance Motor Company.** Fred Paige was a respected businessman who had already gone far in the Detroit business world in the insurance field, where he became a part owner of the Hartford Insurance Company. He also was part owner of a paper company, even though he had three different partners over a matter of just a few years. Despite being unable to get along with his partners, Paige was recognized and respected in the Detroit community as a successful businessman and bookkeeper.

Paige Motor Company factory in Detroit

When the newly formed **Reliance Motor Company** was having financial troubles, they drafted Fred Paige to head the company and steer it toward solvency. The company had great credentials, with a car engine designed by the renowned Henry Leland.

In its second year of business, **Reliance** manufactured a truck which soon outstripped the auto in sales. The board decided to produce trucks exclusively and sell off its auto division. Unfortunately, since Paige had been unable to turn **Reliance** around financially, he was let go in 1907. The truck division was successful enough to later be absorbed by **General Motors**.

Fred Paige found himself in between jobs and decided to build his own car. Starting the **Paige-Detroit Motor Car Company,** the first vehicle produced was the 1909 *Paige.* This was when Henry Jewett entered the picture. Jewett was what might be called a coal tycoon, earning a fortune by bringing West Virginia coal to Michigan. Looking for a place to invest his money, he took a test ride in Fred Paige's three-cylinder, 25-horsepower 1909 vehicle and judged it a good investment. He invested $100,000 in the venture, a formidable sum at the time. However, as he learned more about vehicles, he reevaluated the Paige vehicle and deemed it "a piece of junk!" Jewett eased Paige out as president of the company and took over himself. He closed and retooled the factory. In 1911, the new car that emerged was called the *Paige,* dropping the "Detroit" from its name. This was curious since Fred Paige was gone.

The new *Paige* was a four-cylinder vehicle, and, by 1915, a six-cylinder vehicle was produced. The auto's six-cylinder engine was so popular that by 1916, the company used them exclusively. The *Paige* advertising continually trumpeted its vehicle as "The Most Beautiful Car in America." The *Paige* brand continued to sell well, with steady production throughout the 1920s.

1920 Paige

Early Graham autos

The **Graham Brothers**, Robert, Joseph, and Ray, got their start in the bottling business. The techniques they developed revolutionized the industry. They next turned their attention to the burgeoning auto industry and first developed a kit that could convert a Model T, the most popular car on the road, into a truck. Upon the success of their kit, the brothers turned their attention to manufacturing their own trucks. They had plants in Stockton and Detroit. They used *Dodge* engines, and **Dodge** dealers began to sell their trucks also.

Graham-Paige plant repurposed as a car dealer in the 1970s.
Presently, it is a food factory.

The **Dodge Brothers** bought the **Graham Brothers** truck business and incorporated it into **Dodge** as their truck division. Per their agreement with the **Dodge Brothers**, the **Graham Brothers** promised not to produce trucks, but they didn't make any promises about regular automobiles and began to scout around for an auto company to buy. They found one in **Paige Motor Company**. **Paige** had been around for a while and had a few factories of their own in Detroit. They had started, like many other auto companies, in the Milwaukee Junction section of the city.

The Grahams kept the *Paige* name for the company and became the **Graham-Paige Motors Corporation**. In 1928, the newly restructured **Graham-Paige Motors Corporation** had a lavish opening celebration at the New York Auto Show. Heavyweight champ Gene Tunney and famed football coach Knute Rockne helped unveil the new **Graham** car line. One of the more famous **Graham** models was the 1938 Graham custom sedan, known as the *Sharknose* for its wide front grill and European styling.

Sharknose Graham

Grahams were respected autos, and the **Graham** logo graced cars all through the 1930s. Finally, after the **Graham-Paige Motors Corporation** helped the war effort, after World War II, they decided they didn't want to produce cars anymore. The company was sold to Joe Frazer and Henry Kaiser, and the company morphed into the **Kaiser-Frazer Motor Company**.

A Dodge-Graham dealer in Ann Arbor

The 1930 Graham

The *Carter* (not to be confused with the *Cartercar*) was the first car that Byron J. Carter manufactured. In 1901, in Grand Rapids, a consortium of businessmen gathered to manufacture the car, moving to Jackson later in the year to build it.

The car lasted a year, and Byron Carter then worked for the **Jackson Automobile Company** until 1905. When his new company wouldn't accept his friction-drive transmission, he formed his own company, the **Motorcar Company**, later changed to the **Cartercar Company**, to build the *Cartercar*. They made a quick move to Detroit and then another one to the factory of **Pontiac Spring and Wagon Works** to build *Cartercars*.

A Cartercar

Sales of the popular car, which had variable speeds due to the friction-drive transmission using a large-drive wheel, steadily rose from over one hundred in 1906 to over three hundred in 1908. In 1909, the firm was purchased by Billy Durant for **General Motors.** **GM** manufactured the *Cartercar,* increasing the engine to four cylinders.

The Cartercar factory

Durant was ousted by the **GM** board at the end of 1910. By the time Durant regained control of the company in 1915, the *Cartercar* had been discontinued, and the **Pontiac** factory was used to make the *Oakland,* later renamed the *Pontiac.*

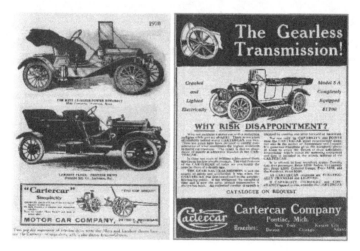

Cartercar dealer and Cartercar ads

Jonesville, Michigan, was where the **Deal Buggy Company** had been building carriages since 1865. George Deal first put a motor on one of the buggies in 1905. When it proved to be popular, he made a few for some of the other locals.

In 1908, the **Deal Motor Vehicle Company** was formed to build a four-cylinder, 30-horsepower auto with a 102-inch wheelbase in runabout, surrey, and touring styles. When designer George Deal died in late 1908, the company's days were numbered. Although they continued to produce autos through 1911 under new head George Dickerson, Dickerson left the company at the end of 1913, and by 1915, the **Deal** company was out of business. The factory was next used by a toy company.

The **Demotcar Company** was a Detroit company—the name was a contraction of "Detroit Motor Car." The car was known as the *Demotcar* and, according to the company brochures, as the *Demot*. The car was a two-cylinder, 10-horsepower runabout that sold for $550 in 1909. The company was started by C. H. Ritter, a wholesale liquor distributor, and the board composed mostly of his friends, who tended to be clueless about auto production.

The Demot or Demotcar

Bankrupt in 1910, the company was once more in C. H. Ritter's hands in 1911. He moved it to Madison, Wisconsin, reorganized the company as **Ritter Automobile Company**, and produced an auto like the *Demotcar,* except it had four cylinders and sold for $685. It was named the *Ritter*. It had about the same amount of success, or lack of success, as the **Demotcar Company**.

In 1909, the **Clarke-Carter Automobile Company** was organized and manufactured the *Cutting,* said to be a "well-thought-out and well-built automobile." It was manufactured in the same Jackson, Michigan, factory that the *C.V.I.* had been built in. Like the *C.V.I.,* the *Cutting* had been designed by engineer Charles Cutting. Instead of six cylinders like the *C.V.I.,* the *Cutting* was a four-cylinder, 35-horsepower vehicle, available in runabout and touring models. Selling for $1,200, the autos made a good showing in the 1911 and 1912 Indianapolis 500s.

1912 Cutting Roadster

Unfortunately, the company was undercapitalized. By 1913, the liabilities of the company topped $350,000, and they went into bankruptcy. The remnants of the company were purchased for $30,000 and the assumption of two mortgages. The buyer, L. C. Erbes, assembled the last three hundred cars in the **Mason** factory in Iowa, and the cars were sold mainly to Iowans.

The car company known as **E-M-F** might be compared to a "supergroup" like rock's Crosby, Stills, and Nash or Emerson, Lake, and Palmer or Major League Baseball's Tinkers-Evers-Chance. Each of the members in the name of the company had already made their mark in auto history, and each would continue after the company was dispersed.

The first major auto body manufacturer was Barney Everitt.

The first successful car dealership and first auto show were originated by William Metzger.

An inventor of machines and mass-production methods, who showed Henry Ford how to improve the assembly line, was Walter Flanders.

Barney Everitt started as a carriage maker in the 1890s. In 1899, he opened his own carriage-trimming business at the corner of Brush and Woodbridge Streets in downtown Detroit. He branched out into making bodies for *Oldsmobiles*. Work became so plentiful that he invited the brothers of the Fisher family of Ohio to make the trek north and help out. They formed the **Fisher Body Company**.

Everitt's auto body business expanded, and he hired Walter Briggs to be in charge. Briggs would go on to be a large body shop builder on his own with **Briggs Manufacturing Company** as well as owning the Detroit Tigers for many years. Everitt was on hand for Henry Ford's early auto experiments and became a principal stockholder and board member of the **Wayne Motor Company**.

Bill Metzger was the father of the auto dealership, started the first Detroit Auto Show, and was paramount in the formation of five different auto companies. He started with the **Northern Manufacturing Company** in 1900.

Walter Flanders was a whiz with machines and, after forming his own machine shops, traveled to the new auto firms, teaching the auto pioneers how to use them. He signed on with Henry Ford and increased his production from twenty to 150 cars per day by rearranging the machines within the factory and teaching the employees how best to use them. He then moved on to teach his production techniques to the **Wayne Automobile Company**. From there, Flanders joined with Everitt and Metzger to form their own company, **E-M-F Motor Company**. The **Northern** and **Wayne Auto Companies** were absorbed into the new company.

The reputations of the three founders were widely known in the auto world of the time, and the first offering of the new company was widely anticipated. The first car to roll off the Flanders-constructed **E-M-F** lines was the 1909 *E-M-F 30*. With a distribution deal with *Studebaker* cars to use their sales network, the new car quickly caused the fledgling company to jump to number two in sales in the United States, right behind the **Ford Motor Company.**

E-M-F continued to be a dominant force in the auto field for the next few years until it was weakened by its agreement with **Studebaker Corporation** enough to be taken over by the J.P. Morgan firm, who also came to end the Studebaker family's dominance over their namesake company. The **Studebaker** name was at first added to the **E-M-F** nameplates, and by 1911, the **E-M-F** name was gone.

1909 E-M-F Touring Car

Everitt formed his own company, **Everitt Motor Company.** Metzger joined with him, and the company name was changed to

Metzger Motor Company. Flanders then joined, and the name was once more changed to give Flanders his turn. The new **Flanders Motor Company** produced cars for several years and was eventually absorbed by the **Maxwell Motor Company**, which in turn became **Chrysler**.

B.F. Everitt Co.   *Metzger Motor Yard*

The E-M-F team made one last appearance on the auto scene together when they aligned with Eddie Rickenbacker to help manufacture his acclaimed car, the *Rickenbacker*, from 1921 to 1926.

Barney Everitt, William Metzger, and Walter Flanders

E-M-F Company factory

Eddie Rickenbacker was a heroic pilot in World War I, fighting the Red Baron and surviving many ordeals in the Pacific war theater. After the war and parades, Rickenbacker pursued several livelihoods, and making autos was one of them. Starting off with some of the best auto minds in the business with the **E-M-F** crew, he got famed auto factory and skyscraper designer Albert Kahn to design an auto factory that was built on 4815 Chabot Street in Detroit.

Rickenbacker autos

The Chabot Street, Detroit factory of the Rickenbacker Auto Company

The *Rickenbacker* was a sturdy, well-built car that stayed in production for seven years, a long time for the day. It was available in sedan, coupe, and touring styles and sold for $1,500. The car was also a pioneer in introducing four-wheel brakes (along with the *Wills Sainte Claire*) on their cars.

Eddie Rickenbacker bowed out a year before the company went out of business. After the **Rickenbacker Motor Company**, he purchased the Indianapolis Speedway, where the Indianapolis 500 was run each year. After that, he founded his own airline, Eastern Airlines.

One of Henry Ford's earliest investors was Alexander Malcomson, a businessman who made millions as a coal merchant. He owned a little over 25 percent of the company in the early days.

When he saw Henry Ford not making any moves to have the company produce anything other than the *Model T*, Malcomson decided to venture out on his own. He built his own factory at Mack and Beaufait Streets in Detroit to produce his own automobile, a luxury car he called the *Aerocar*.

In 1906, Malcomson's fledgling car company manufactured a four-cylinder, 24-horsepower auto in the new factory. The *Aerocar* had the motto "Practical Cars Built by Practical Men," but by 1908, they were out of business.

Upon the failure of his company, Malcomson needed money and had to sell his **Ford** shares to get out of financial trouble. Henry Ford was only too glad to buy back the shares so he could consolidate his control of the company. Malcomson quit the auto business, got back into coal, and was once again a millionaire by the time he died.

One of the many notable automobiles of Detroit was the *Scripps-Booth*, which was manufactured by a company that operated and manufactured autos out of Detroit for ten years. The company founder's grandfather was James E. Scripps, who was born in London, England, moved to Rushville, Illinois, and then started working at the *Chicago Herald* newspaper. He got transferred to the *Detroit Herald* and became manager of that newspaper as well as part owner of the *Detroit Daily Advertiser*.

The *Detroit Daily Advertiser* offices burned down, and Scripps received $20,000 of fire insurance money, which he used to start his own newspaper, the *Evening Daily News*. This paper changed its name to the *Detroit News*. Following his success in Detroit, he formed a newspaper dynasty with his family, acquiring and starting newspapers in Cincinnati, Cleveland, St. Louis, and Chicago.

The philanthropy of James Scripps and his family included major funding for the Detroit Institute of Arts, Cranbrook Institute, and much more. By the third generation, the money was still holding out, and grandson James Scripps-Booth started his own automobile company in the Motor City. Besides starting the car company, Scripps-Booth was a renowned artist, with works hanging in many famous galleries.

For ten years, from 1913 to 1923, the **Scripps-Booth Company** put out a cyclecar, the *Bi-Autogo*, which had one of the first V-8 engines. The company also produced luxury cars until 1923, when the company was purchased by **General Motors**.

Scripps-Booth cars, 1916 Model Cs, and an earlier model. The three-wheeled vehicle is the Bi-Autogo.

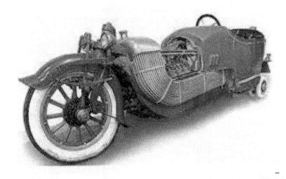

The **Standard Electric Auto Company** had a four-year run from 1911 to 1915, manufacturing electric cars in Jackson, Michigan. Although the *Standard Electric* could only go a top speed of twenty miles per hour, it could go for 110 miles on a single charge, good for an electric car in the 1910s.

After an infusion of capital, the company was reorganized as the **Standard Car Manufacturing Company** in 1913. After lawsuits where the plaintiffs were accused of suing just to get back at the company officers, the company went into receivership in 1915.

1912 Standard Electric of Jackson, Michigan

The *Argo Electric* was built by the **Argo Electric Vehicle Company** of Saginaw, Michigan, from 1912 to 1914. It used a 60-volt system with Westinghouse engines and steering wheels (not tillers). Four- and five-seaters, in open and closed models, were capable of going twenty miles per hour and were a very good deal, according to the manufacturer. Unfortunately, the public didn't agree.

1913 Argo Electric Fore-Drive Limo Model

The *Argo* name was popular, as the **Argo Motor Company** produced a gasoline vehicle from 1916 to 1918 in Jackson, Michigan. It had no relation to the Saginaw *Argo*. The Jackson *Argo* was a cyclecar, which was one of many produced that were inexpensive, one- or two-cylinder cars that were usually narrow, seating only two or three people. The founders were Ben and Frank Briscoe, and they took over the factory of the **Standard Electric Car Company**, building a four-cylinder, 12-horsepower cyclecar called the *Motorvique* until 1917. It had a sliding-gear transmission, shaft drive, and ninety-inch wheelbase, weighing 750 pounds and selling for $295. In 1916, with the end of the cyclecar trend, a larger car with a 22-horsepower engine, called the *Wego*, was assembled.

The Standard Electric Car Company factory,
later used by the Jackson Automobile Company

The **Krit Motor Car Company** was started in Detroit in the summer of 1909 to manufacture the cars designed by Kenneth Crittendon, who was previously with **Regal** and **Ford**. The first autos were released in 1910 with a four-cylinder engine and won a number of hill-climbing contests.

A hardy car, it was popular among farmers in the West, which is why the drought of 1913 was said to have affected sales. Also in 1913, the company released their only six-cylinder car, the *Krit Six*.

The Krit radiator badge

At first, the car was spelled *K-R-I-T* but was simplified to *Krit*. The **Krit** company had a swastika for their car symbol (twenty-five years before Hitler), which they said was a symbol that meant "to ensure favor of auspicious gods." It only worked a few years, though, because the company was usually in financial straits, switching owners a few times. They also went from the old **Blomstrom** factory and the old **Owen** factory. They went bankrupt in 1915.

Krit factory in Detroit

The **Hackett Motor Car Company** was located in both Jackson and Grand Rapids. Starting as the **Disco Starter Company of Detroit**, founder Mansell Hackett bought the **Argo Motor Company** of Jackson from the Briscoe brothers in 1917. They manufactured four-cylinder, rotary-valve engine autos named *Hackett*s as well as assembling *Argo*s until 1918, when the parts ran out. The *Hackett* was available in a five-passenger touring car and a four-passenger roadster and priced at $888. The *Hackett* had a four-cylinder G.B. & S. engine.

Due to World War I, it was difficult to get parts, and the **Hackett Motor Car Company** was idle for part of 1918. During this time, they moved their factory operations to Grand Rapids, Michigan, and resumed operations in 1919, but it was too late, and Hackett dissolved the company at the end of 1919.

The Hackett Factory (photo courtesy of american-automobiles.com)

The total production of *Hackett*s had been about 118 cars. Mansell Hackett went back to selling auto starters, and the *Hackett*'s rotary-valve engine was taken by its developer Fred Guy to the **Apex Motor Company** in Ypsilanti for the auto called the *Ace*. The auto factory in Grand Rapids was used to assemble the *Lorraine* auto after Hackett left.

In 1922, **Commonwealth Motors** of Joliet, Illinois, and the **Markin Body Corporation** merged and formed the **Checkers Motor Corporation**, manufacturing *Checker* cabs until well into the 1980s in Kalamazoo, Michigan.

Checker cab

William Durant bought the **Welch-Detroit Motor Company** and the **Rainier Motor Company** of Saginaw. In 1909, he formed the **Marquette Motor Company** in Saginaw to continue producing the *Rainier* and parts for the *Welch*. Both cars stayed in production until 1911. A marque called the *Marquette-Buick* was built in Saginaw to be a race car.

After Durant exited **General Motors** in 1910, the company decided to consolidate *Rainier* and *Welch* and produce the 1912 *Marquette*, a four-cylinder, 40-horsepower, touring-style auto. **GM** discontinued the marque after 1912.

Marquette Motor Car Company, Saginaw factory

Hugh Chalmers was vice president of the **National Cash Register Company** and decided that he wanted to get into the automobile business. In 1907, he purchased half of the **Thomas-Detroit Company.** This was the company that was originally spun off from the **E.R. Thomas Motor Company** of Buffalo, New York, the makers of the *Thomas Flyer*. In 1908, he renamed the company the **Chalmers-Detroit Motor Company** and kept on producing basically the same car that **Thomas-Detroit** had manufactured.

By 1909, Chalmers had purchased most of the company and renamed it the **Chalmers Motor Company**, dropping "Detroit" from its name. The cars were popular six-cylinder cars, available in all styles and models, including phaeton, tonneau, touring, limo, and more. They were well known for giving out the Chalmers Award each year to the baseball player "most important and useful to his team." Ty Cobb was the first winner. Hugh Chalmers was piqued when Cobb sold the car given to the winner, but he continued to give the award until the **Chalmers Motor Company** merged with the **Maxwell Motor Company.** The last *Chalmers* auto was built in 1924 and was released in 1925. The **Maxwell Motor Company** went on for one more year before it was transformed by Walter Chrysler into the **Chrysler Motor Company**.

Chalmers factory pictures, ad, and a 1922 Chalmers

Detroit had its share of unusual car and car company names. Perhaps none was as unusual as the *DODO*, which was a Detroit cyclecar manufactured in 1914, the big year for cyclecars. It was a two-seater automobile, larger than most cyclecars, with a narrow, one-hundred-inch wheelbase and a two-cylinder, 10–12-horsepower, air-cooled engine, built by the **Auto Parts Manufacturing Company of Detroit** and designed by engineer Karl Probst. It featured an early front-wheel drive. Problems ensued, and the *DODO* proved true to its name and became extinct. Probst later worked on the *Bantam* and the *Jeep*.

DOLSON

In Charlotte, Michigan, **John L. Dolson & Sons** had been making carriages for many years and had production figures of over five thousand a year. In 1902, son Elmore started experimenting with autos and, by 1904, had manufactured a two-cylinder touring car. In 1905, the firm brought out the *Cannon Ball Roadster,* a four-cylinder car which could go in excess of seventy-five miles per hour.

The 1905 Dolson of Charlotte, Michigan

In 1906, the name of the company was changed from **J.L. Dol-**

son & Sons to the **Dolson Motor Company.** The car was called the "Durable Dolson" and was known as a well-made car of wood and steel. It had 60 horsepower under the hood and was advertised as the "Mile-A-Minute" car. In 1907, their seven-seater touring model sold for the princely sum of $3,250. The Dolsons also offered smaller cars: shaft- and chain-driven, two-cylinder ("flat-twins"), 20-horsepower models and a shaft-driven, four-cylinder, 30-horsepower model.

Unfortunately, in 1908, the company was accused of selling more stock than it was allowed. They went into receivership, and in 1909, the factory was sold to the truck-producing **Duplex Power Car Company.** The father, John Dolson, then went into peppermint farming, and the sons stayed working in the automotive field in sales and parts.

The **Hudson Motor Company** was one of the most famous and most beloved of the now defunct automobile companies. Maybe this is because of the *Cars* movie by Pixel Productions, where a

Hudson Hornet was one of the main characters. Its popularity was such that even before the movie, Hudson Clubs could be founded all over the United States and Canada after the company's exit from the auto stage.

The **Hudson Motor Company** began when Roy Chapin, Howard Coffin, and Roscoe Jackson met at the Phi Delta Theta fraternity house at the University of Michigan. Chapin took a ride in an *Oldsmobile* and decided that making cars was what he wanted to do. He got his friends Howard and Roscoe interested in designing autos and forming an auto company. They designed autos for the **E.R. Thomas Company** and the **Chalmers Motor Company** before they joined the **Hudson Motor Company**.

The first car designed was the *Hudson Twenty*, which achieved the goal of Chapin and friends to design a reliable car for under $1,000. It was built in the factory at Mack Avenue and Beaufait Street in Detroit that was formerly used to build the *Aerocar*. The *Hudson Twenty* was successful, propelling the new company to continue manufacturing automobiles. They eventually became the number-three automaker behind **General Motors** and **Ford**, over-taking **Nash, Packard, Cole, Franklin**, and the myriad other auto companies producing autos then.

In 1913, the **Hudson Motor Company** moved to a large, Albert Kahn-designed factory at Jefferson Avenue and Connor Street in Detroit. The year 1916 saw the *Hudson Super Six* come out (it had six cylinders) and set a new land speed record of 102.5 miles per hour. Later the same year, it set a record for climbing Colorado's Pikes Peak in 18 minutes and 25 seconds. The *Super Six* kept **Hudson** a top-selling company for over a decade.

The first Hudson, Earhart with Roy Chapin and other Hudson executives

Amelia Earhart ad for the Terraplane

The Hudson Hornet, Hudson trading card

In the early 1900s, the bicycle craze was beginning to give way to automobiles. As auto technology improved and car culture spread, more people bought autos. A surprising amount of people who purchased autos thought they could manufacture one better than the one they had purchased.

1936 Terraplane

After the auto craze at the turn of the twentieth century, an aircraft craze began. The Wright Brothers and World War I ushered in aircraft engine research and advancements. The auto business recognized this; the **Hudson Motor Company** had a car called the *Terraplane,* and they advertised it by saying that it felt like "land flying." They even hired famed aviatrix Amelia Earhart to publicize it.

The company later achieved success with the *Special Six,* the *Essex,* and the *Wasp* besides the *Terraplane.* Surviving the Great Depression, the **Hudson** factories converted to war production during World War II.

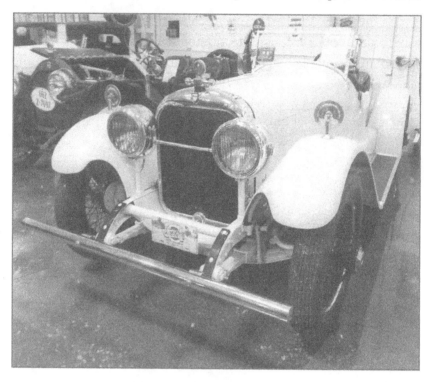

Hudson Essex

After the war, in 1946, **Hudson** brought out the *Hudson Hornet*, a very successful model, but by the 1950s, it was becoming harder for the smaller car companies to survive. When Hudson brought out the less-than-successful, compact car the *Jet* and the sporty *Italia*, which lost a lot of money, they decided to merge with the **Nash-Kelvinator Company**. They became the **American Motor Company (AMC)** in 1954. The **Hudson** nameplate was used until 1957. **AMC** was purchased by **Chrysler** in 1986.

The last Hudson dealer in the United States in Ypsilanti, Michigan, and the first Hudson factory at Mack and Beaufait

The Hudson factory at Jefferson Avenue and Conner Street (Detroit)

Hudson factory, Conner Avenue view

The **Continental Motors Company** started off in Grand Rapids, Michigan, as an engine company into the largest engine supplier in the United States. From Grand Rapids, they moved to a factory on Jefferson Avenue in Detroit, near the **Hudson** factory.

In 1935, when the **De Vaux-Hall Motor Company** failed, the **Continental Motors Company** manufactured the six-cylinder roadster that had been the *De Vaux* as the *Continental*. When that venture failed in 1936, they went back to just making engines.

Continental Motors in Detroit

The Dodge brothers were as famous for their partying ways as they were for their trendsetting cars. Starting out poor in Niles, Michigan, they went on to work in machine shops and then started manufacturing parts for the **Olds Motor Company** and then the **Ford Motor Company**. They were one of the earliest shareholders of **Ford** stock.

Henry Ford wanted to phase out his distributors, like the Dodge brothers, and make all of his parts in-house. The *Model T* was inexpensive but well-made, and as long as the *Model T* was all they produced, Ford could pull it off. It's no surprise that Ford was upset when his engineering department made a prototype for a *Model T* successor. He went out and smashed the prototype! The Dodge brothers, already having been pressured by Ford to sell their stock and losing the **Ford** parts business, took the prototype, refined it, and produced the first successful Dodge brothers car. They went on to produce successful autos for many more years.

The brothers were not ones to be all work and no play. They would take a liberal approach to drinking in the workplace, especially when they would find their workers at the bar drinking instead of working. They would buy a keg and bring it into the plant to keep the workers working. When the Dodge brothers were in celebratory mode, it was usually done in out-of-the-way dive bars. They would generally trash the place, much like a rock star in a hotel room, and then they'd go back the next day and pay for all the damages, although admittedly, sometimes, they partied so much that they couldn't remember where it was.

Many stories abound about how Horace or John would dance on the bar and throw shot glasses at the mirror or throw them at the feet of the proprietor while HE was made to dance on the bar. Polite society frowned on the rich Dodge brothers, so when the Grosse Pointe Yacht Club wouldn't let Horace join, he bought the land next to the sumptuous club and built a mansion that dwarfed the yacht club in both size and elegance. Even in death, they were flamboyant. The Dodge brothers are buried in an Egyptian-motif mausoleum in Detroit's Woodlawn Cemetery.

The Dodge radiator emblem, the 1914 Dodge

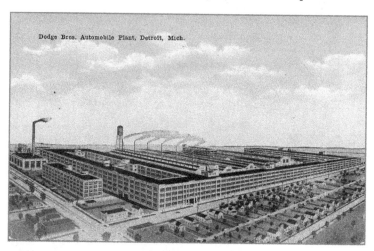

Dodge Main in Hamtramck, Michigan

In 1911, the **Miller Car Company** was organized in Detroit by Theodore Miller. It was housed within the factory of the **Detroit Excelsior Works**. The different models included a four-cylinder, 30-horsepower roadster, mounted on a 110-inch wheelbase. It sold for $1,250. Another model was a touring car with a 116-inch wheelbase, selling for $1,350. In 1913, a one-thousand-pound delivery truck was added to the offerings. When the **Miller Car Company** went into receivership in 1914, the **Kosmuth Company** stepped in and continued to make the delivery truck. The *Miller* design was incorporated into the *Pennsy* of Pittsburgh.

The 1912 Miller

In 1910, the **Wolverine Motor Car Company** formed in Mount Clemens, the county seat of Macomb County, Michigan. A 30-horsepower, four-cylinder, gasoline-powered prototype was produced but never built in Mount Clemens. Instead, Thomas F. Ahern bought the **Wolverine** equipment and machinery and moved everything north to Alpena, Michigan. This was because the Alpena Chamber of Commerce offered Ahern $50,000 to get the town into the automobile business. Forming the **Alpena Motor Car Company**, Ahern used the **Wolverine Motor Car Company** machinery to build the four-cylinder *Alpena Flyer.*

Wolverine Motor Company factory

The 1910 New York Auto Show put the prototype *Wolverine/ Alpena Flyer* in the Top Ten Best Autos! The car had a water-cooled engine and a sliding-gear transmission with three speeds and reverse and a disc clutch and was only available in royal blue.

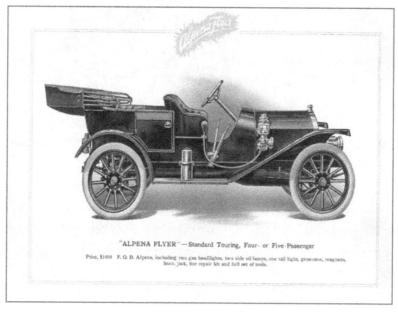

"ALPENA FLYER" — Standard Touring, Four- or Five-Passenger

Price, $1450 F. O. B. Alpena, including two gas headlights, two side oil lamps, one tail light, generator, magneto, horn, jack, tire repair kit and full set of tools.

Alpena Flyer auto and factory photos (photos courtesy of the Besser Museum, Alpena, Michigan)

In 1913, "Flyer" was dropped from the name, and the car became known as just the *Alpena*. In 1913, they manufactured vehicles with six-cylinder, 50-horsepower engines and a 135-inch wheelbase, but it was too little too late. By early 1914, the company was hemorrhaging money and went into receivership. The **Besser Manufacturing Company** bought the assets, including the factory. The *Alpena Flyer* may be gone, but to this day, the **Besser Manufacturing Company** is still in Alpena.

The Alpena Auto Factory

The Fuller brothers, Charles and Frank, first approached the Blood brothers, Maurice and Charles, in 1902 to help them build an auto. This was because of the wide renown of the Blood brothers' mechanical skills.

The Blood brothers started out making bicycles in Kalamazoo, Michigan, and that's probably why their first auto only had a forty-eight-inch wheelbase, which is a tiny car. The next prototype had a fifty-four-inch wheelbase, still a small machine, but this was the first one that they manufactured to sell. They called it the *Michigan*. The company name was the **Michigan Automobile Company, Ltd.**, and the early cars were all called *Michigan*s, but in 1905, the Bloods and the Fullers quarreled and parted ways.

The Michigan auto

The Bloods continued making autos and renamed their car company the **Blood Brothers Automobile & Machinery Company** while the Fullers continued to make the same cars under the *Michigan* name. The cars all had two cylinders, and the *Model C* was 12 horsepower with a seventy-eight-inch wheelbase. The *Model D* was a light touring model with 12 horsepower and an eighty-inch wheelbase, and the *Model E* was a side-entrance tonneau model with 16 horsepower and a ninety-inch wheelbase.

By 1906, the Blood brothers quit making autos and dropped the "Automobile" portion of their name, making only universal joints, but they started up again in 1914 and 1915 as the **Blood Brothers Machine Company**, manufacturing a cyclecar they named the *Cornelian*. The *Cornelian* was a four-cylinder cyclecar with independent rear suspension, rare on a car in 1914. A racing version was driven by Louis Chevrolet in the Indianapolis 500 of 1915.

Racing version of the Cornelian driven in the
1915 Indy 500 by Louis Chevrolet

The Blood brothers found that making just universal joints was more profitable than auto production and dropped the autos (again) in 1916.

The **Michigan Buggy Company** of Kalamazoo, after selling buggies for thirty years, branched out into auto production with the *Michigan*, an attempt to build a luxury car for a low price. That price turned out to be about $1,150 to $5,000 for the top model, known as the *Mighty Michigan,* which had 40 horsepower, double that of the **Ford** *Model T.* The less expensive *Model D* was also an impressive 33 horsepower.

The auto company was made a separate division and titled the **Michigan Motor Car Company**. They successfully produced cars until 1915, when a payroll scandal shattered the company.

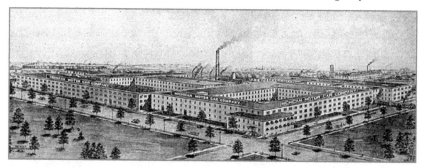

This first rendering of the Michigan Motor Car Factory is typically exaggerated to make the car marque seem more important by making the plant look bigger.

Michigan Motor Car Company logo

The 1913 Michigan Model R touring car

The **Brush Motor Car Company** started in 1907 and changed its name to the **Brush Runabout Company** in 1910. The car had a one-cylinder, 6-horsepower engine and was known for its hill-climbing abilities, selling in the $500 range. Alanson Brush designed the car and had also helped to develop the *Cadillac* and the *Oakland* (later *Pontiac*) automobiles.

1912 Brush Runabout

In 1910, the **Brush Runabout Company** became one of the companies in the Briscoe brothers' **United States Motor Company** along with the *Stoddard-Dayton, Maxwell,* and *Columbia* autos and a few others. When the company went under, only the *Maxwell* survived. The last *Brush* was manufactured in 1912.

1911 Brush Panel Truck (on trading card)

1909 Brush Runabout

The Brush factory in Detroit

In late 1911, Claude Briggs of the **Brush Runabout Company** and John Boyle combined to form the **Briggs-Detroiter Company**. Their goal was to make a smaller, less expensive version of the popular *Brush Runabout*. The car, called the *Detroiter,* at first had a Continental four-cylinder engine, then offered a Perkins V-8 engine, and then a Continental six-cylinder engine.

Business was brisk at first, but as sales slowed in 1915, the company went into receivership. Reorganizing, they produced autos until 1917, when the assets were sold at auction.

In 1900, the **Chelsea Manufacturing Company** made small, metal novelties for the boss, A. R. Welch, in Chelsea, Michigan. In addition, A. R. and his brother Fred had a bicycle business together. In 1901, the Welch brothers built their first auto, including designing their own engine, a two-cylinder, 20-horsepower unit that had overhead valves.

In 1903, production began on the auto, and soon, the **Chelsea Manufacturing Company** was in the auto business and manufacturing about fifteen *Welch Tourist* cars per month, but 1904 brought about bankruptcy, and production ended for a short time.

Reorganizing as the **Welch Motor Car Company** and acquiring $100,000 in capital, the Welch brothers moved to Pontiac and began manufacturing a larger version of the *Welch Tourist.* It was one of the largest cars built at the time. When being exhibited at the Detroit Car Show, it was too big to go through the entrance and had to be disassembled and reassembled inside.

By 1905, the company was manufacturing four-cylinder autos with 36 horsepower. By 1907, a six-cylinder auto with 75 horsepower was being produced. The car had a 138-inch wheelbase. In 1908, **Welch** acquired a second factory in Detroit to make a smaller version of the car called the *Welch-Detroit* and continued making large autos in Pontiac. They sold a $4,500 *Model 4* touring car and a $7,000 *Model 6* limousine. In 1910, they started to be associated with **General Motors** and by 1911 had been purchased by **GM**.

A. R. Welch had plans to manufacture a new cyclecar design he had but was unfortunately never found after he disappeared in 1913 following a duck-hunting trip. His empty canoe was all that was recovered.

1904 Welch Tourist

The *Day Utility* was manufactured on Trumbull Street in Detroit by the **Day Automobile Company** (headed by Thomas Day) from 1911 to 1913. It was a four-cylinder, 21-horsepower (26-horsepower in 1912) vehicle touted to farmers by advertising, "All the room possible is provided for carrying produce." This was an area scooped out under the two seats.

At first, the car sold well enough for the company to move to a Milwaukee Junction factory at 25 E. Milwaukee Street, but sales slowed, and in 1913, Thomas Day said that he had secured further financing in Spokane, Washington, and would be moving the whole operation there, but it never happened.

The **Detroit Novelty Machine Works Company** became the **Detroit Vehicle Company** in 1904 and was capitalized at $150,000. Having enough money was paramount in the fortunes of early car companies and whether they would make it longer than their first year. Many, if not the majority, of companies that were formed to build cars either built a prototype only or, more likely, no car at all. That was the fate of many companies that tried to start with less than $50,000.

The most famous investor was Joseph L. Hudson, owner of the J.L. Hudson Department Store. This was his first car venture; his second was more successful and named after him, the **Hudson Motor Company**.

The first manufactured vehicle of the **Detroit Vehicle Company**, also known as the **Detroit Auto Vehicle Company**, was the 1905 *Crown Side Entrance Tonneau* auto, which was a 24-horsepower, five-passenger touring car. Also in 1905, the company had a 12-horsepower *Crown Runabout* that sold for $750. They also manufactured a 16-horsepower delivery vehicle. Besides the company's factory at 42 W. Larned Street in Detroit, the company also had a foundry in Romeo, Michigan. The cars had some problems, and the 1906 version was redesigned by Edward T. Ross of the **Cadillac Motor Company**.

The 1907 auto line included a two-cylinder, 24-horsepower engine which the company claimed ran as smooth as a four-cylinder or six-cylinder engine. That year's models included a runabout for $1,500 and a touring car for $1,600. These models were called *Crown-Detroit*s or just the *Detroit*. The entire 1907 output was purchased by car dealer and future automaker John Willys.

Despite the success of selling out all the 1907 vehicles, many stockholders of the original company, the **Detroit Novelty Machine Works Company**, objected to the change to auto manufacturing and had been fighting the new company since its inception. Their litigation brought the company to bankruptcy at the end of 1907. The bondholders of the company received $24 for every $100 they invested.

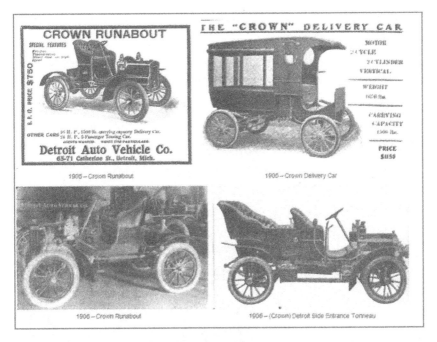

The Crown Motor Company

At the turn of the nineteenth century to the twentieth century, lots of men figured that the road to riches lay in starting their own auto company. Similar to the Internet boom of the twenty-first century, for some, it was, but for many others, not so much. Many car makers were able to plug along for many years, not really making it but hanging in there enough to manufacture a new model for the next year, always hoping that it would be the one that would take off in a big way.

Homer Warren was a former postmaster of Detroit and obtained financing in 1909 to start a factory. He built the large, two-story, brick building at 1331 Holden Street in Detroit, near the Midtown area. He also hired technicians and administrators from the **Hudson** and **Olds Motor Companies** and manufactured the first car of the **Warren Motor Company**, the 1910 *Warren*. It was an auto with a 30-horsepower, four-cylinder engine. This was the specification for most of the cars that were manufactured by **Warren**.

The company fielded EIGHT different models in its next year, 1911. These included touring, torpedo, and "dickey roadster" models. In 1912, it was down to a roadster and a few others. Financial problems ensued, and Homer Warren got an extended line of credit for the following year. In 1913, a six-cylinder model, the *Resolute*, was added. The other two models were the *Pilgrim* and the *Wolverine*.

The main way to promote the *Warren* was going to be on the racetrack—and the *Warren 30* did go on to win a few races, but ultimately, the lack of an effective car dealership system did them in. Since the *Warren* had no car dealership system of its own, it contracted out to a Philadelphia firm. In 1913, when the Philadelphia sales firm went under, so did the **Warren Motor Company**.

A Warren-Detroit ad

Homer Warren

The factory that Homer Warren built is still standing and has been repurposed into a recycling plant.

Adrian, Michigan, was the place where the **Lion Motor Company** manufactured the *Lion 40* roadster from 1910 to 1912. At first called the **Church Manufacturing Company,** they had earlier (in 1904) attempted to produce the *Murray* and the *Lenawee* autos with little success. They were single-cylinder, $1,000 autos. Only about fifteen were made. The *Murray* was named after Willis Grant Murray, a traveling salesman, who convinced the **Church Manufacturing Company**, a car-parts manufacturer, to go (unsuccessfully) into the car biz.

The **Lion Motor Company** was a little luckier. They managed to manufacture cars for a full two years before their eventual failure. They put out three hundred *Lion*s; however, the demise of the company came when two hundred more were destroyed in a 1912 fire that destroyed the cars and the company's factory and future. One *Lion* and one *Murray* are still left in the Lenawee County (Michigan) Museum.

The 1914 Lion

The *Koppin* was a cyclecar that got its start in Fenton, Michigan, in Genesee County. A cyclecar was often, but not always, a three-wheeled vehicle, considered a cross between an auto and a motorcycle. It usually seated two and was powered by a small engine of just one or two cylinders. Oscar J. Howick had worked for **Lozier** and **Packard Motor Companies** just before working for the **Fenton Cyclecar Company**. It was there that he designed a car called the *Duquesne* in Pittsburgh. He was the designer of the *Fenton*.

The **Fenton Cyclecar Company** was started by auto salesman George Jenks. The *Fenton* had a two-cylinder, 10-horsepower, air-cooled De Luxe engine. With a ninety-six-inch wheelbase, it could easily drive down a sidewalk. Knowing the bad shape of the roads in those days, this was probably a major selling point. The car had a friction-type transmission mounted on large ball bearings. The body was roadster style, and it had a deep vee front with full doors and fenders, an item often scrimped on in cyclecars. The vehicle sat two passengers side by side and sold for $375 in 1913. It was referred to as the *Signet* in a couple of early advertisements, but by the time production began in the fall of 1913, it was named the *Fenton*.

Only about a dozen *Fenton* cyclecars were made before the company's shop went up in flames. Further tragedy struck on March 23, 1914, with the sudden death of company founder George Jenks. However, by May, the company had been reorganized by real estate man Henry Samuel Koppin as the **Koppin Motor Company**. Koppin owned the building known as the A.J. Phillips plant in Fenton, and production was moved to the plant.

The A.J. Philips factory was named after Andrew J. Philips, who later donated his office to the village of Fenton to use as a public library. The business began in 1869 when Philips moved to Fenton and purchased a building at the corner of S. LeRoy and Mill Streets. Here, along with his brothers, the firm conducted business making wooden pumps. They soon branched out into making wooden shovels. In the 1880s, Phillips invented a sliding-window screen, which the factory began to manufacture, hiring fifteen more employees.

In 1883, Philips purchased the lot next door and expanded his factory. He continued to expand through 1886, and the business soon sold all kinds of wooden products, including ironing boards, ladders, clothesline poles, hat and coat racks, benches, lawn swings, cutting boards, bicycle stands, and many other wooden products of the day.

When Phillips died, his sons ran the business until 1913, when the demand for wooden products was decreasing. Closing the business and dividing the assets, Koppin was now in possession of the huge factory. He moved production of the **Koppin Motor Company** vehicle to the complex.

The name of the vehicle was changed to the *Koppin Model A Roadster*, and the price was raised $10 to $385. Advertising also reflected that it now had 13 horsepower and included a top, side curtains, a specially designed windshield, and electric side and tail lights powered by a 6-amp plate storage battery, controlled by a dashboard switch. It also came with a horn, a repair kit, pump, and tools.

The new venture lasted until September 17, 1914, when, at about 1:45 A.M., an explosion occurred, and within minutes, the building burst into flames. Because of a brisk wind, the fire spread to the other buildings. By the time the Fenton Volunteer Fire Department arrived at 3 A.M., the factory was burning out of control. By the time the Flint Fire Department arrived to assist, the factory complex was a total loss. Koppin had been insured and decided that the profits weren't enough to continue producing the vehicle.

Koppin moved back to Detroit, where he got involved in the theater business, owning several famous theaters. The Koppin Theatre on Gratiot Avenue was a famous Paradise Valley blues venue. Other Koppin theaters were located throughout the Detroit area until the 1940s. A black vaudeville circuit was known as the Koppin Vaudeville Circuit.

Oscar Howick went on to design another car in Fenton, the *Elgin Light Car*. This was a 24-horsepower, four-cylinder vehicle, but it apparently never went beyond the prototype stage.

The Fenton Cyclecar, later called the Koppin, was narrow enough to drive on the sidewalk, as the top picture with Mr. Koppin at the wheel demonstrates. The photo on the bottom is the Wayne Theatre, part of the Koppin Theatre circuit.

———————

Benjamin Briscoe caught the auto bug and in 1916 started producing a car in Jackson, Michigan, that he designed himself, the *Briscoe*. It had four cylinders and was later available with eight. They had a special exchange deal if you decided later that you wanted the eight-cylinder model instead of the four—you could have an eight installed for little more than the cost difference between the eight and the four! The only quirk with the first *Briscoes* is that they had only one headlight in the middle of the front grill, which made it illegal in many states.

Briscoe Roadster

Ben Briscoe got tired of running the company and had Clarence Earl doing most of the work. Finally, in 1921, he sold the company to him. Clarence Earl changed the name of the car and the company to his last name. The **Earl Motor Company** got off to a rocky start—they were plagued by pranksters who first ordered and had sent to them ten thousand MENU CARDS!! After a few other pranks, they ordered expensive awnings for all the factory windows! Clarence Earl quit the company after disagreements with the board of directors. The company went on for another three years producing the *Earl*.

1923 Earl, 1921 Earl Touring Car, and an Earl ad

The **Carhartt Automobile Company** of Detroit said in their ads that they had "twenty-eight years of experience" when they brought out their auto in 1911. They didn't mention that the experience was in the manufacturing of overalls. The car that they built in 1911 had a four-cylinder engine with either 25 or 35 horsepower and was available in *eight* different models, including landaulet and limo. In 1912, they trimmed the number of models offered to six but began offering a 60-horsepower engine.

Company founder Herman Carhartt gave up car manufacturing when, at the end of 1912, he was $16,000 in debt. He returned to manufacturing overalls.

William Paterson started his Flint, Michigan, self-named company, the **W.A. Paterson Company,** in 1909 and assembled well-built, well-regarded vehicles until 1923. Starting in the carriage business in 1869, he stopped making carriages in 1910, a year after he entered the auto biz.

The *Paterson* was a vehicle with a two-cylinder, air-cooled engine with a double-chain drive, solid rubber tires, and a planetary transmission. In 1910, the auto gained a four-cylinder engine with 30 horsepower, a shaft drive, and selective transmission. Sales increased from sixty-four cars the first year to 450 cars the next. When in 1915 the company released a six-cylinder vehicle, sales climbed to nine hundred. Four-cylinder cars were eliminated in 1916, and after that, only Continental six-cylinder engines were used.

The art deco onetime headquarters of the W.A. Paterson Company is still in Flint.

The *Paterson* had dealers in all forty-eight states (New Mexico and Arizona having recently joined the Union). The car was a steady seller and had strong brand loyalty. The eighty-three-year-old William Paterson died in 1921, and his son kept the company going through 1923, when it was sold to a **Dodge** dealer, Dallas Winslow, who planned to revive the marque but changed his mind. Over twelve thousand *Patersons* were sold over the life of the company.

Paterson radiator badge; 1917 Paterson ad

Chapter 23:
International Harvester, Luverne,
and More Minnesota Car Companies

Most people rarely think of the state of Minnesota as an early auto center, but it had its own unique auto history starting with Joseph Brown, who constructed an early steam car in 1860 in Henderson, Minnesota.

In 1907, one of the first and most successful motorized vehicle companies in Minnesota, the **International Harvester Company** in St. Paul, decided to branch off into autos. They were more known for tractors and other farm equipment than autos, but in 1909, they produced autos and trucks in the building that the **Willys-Overland Motor Company** vacated.

The 1907 International Harvester

Although the **International Harvester Company (I.H.C.)** had a "motorized buggy" prototype as early as 1899, it wasn't until 1907 that an auto was added to the company's line of tractors and other agricultural equipment. The first one hundred autos, called the *International,* were built in the company's Chicago factory before production was transferred to their Akron, Ohio, factory. The first cars had two-cylinder engines, friction transmission, and solid tires. In 1910, the auto began to be branded with the company initials, **I.H.C.**, and the engine size increased to four cylinders, with a choice of air- or water-cooled engines. In 1911, the company discontinued their auto line but continued to custom-build autos in the 1930s.

From 1903 to 1909, carriage-builder brothers Fenton (Al) and Ed Leicher started the **Luverne Automobile Company** in Luverne, Minnesota, and built cars identified as "high-wheelers" in the style of horse-drawn buggies. It was a common practice when carriage builders started building autos to build them like what they already knew.

Prior to 1903, Al heard a steady customer of the buggy company say that he wouldn't be buying any more buggies because he was going to buy an automobile next. This got Al to thinking that he should check into these new vehicles that looked to be destined to put carriage makers out of business.

With his brother Ed, Al purchased a kit car and assembled it. After modifying the design, they used wheels, hubcaps, springs, fenders, bodies, and hoods from the **A.L. Dyke Automobile Supply Company** in St. Louis, and whatever parts they needed that they didn't get there, they got from the **St. Louis Motor Carriage Company**. Then, they assembled their own original auto.

In 1905, plans were made for a new, $5,000 factory, which started construction and was completed in 1907. A three-story, brick building, it was "well-lighted and had the modern amenities of elevators and other modern appliances," as the *Luverne Herald*, the local newspaper, wrote.

Luverne factory

The Luverne factory, built in 1906–1907.

By 1909, the *Luverne* cars were more conventional and somewhat advanced for the day, including 40-horsepower vehicles with Rutenber, and later Beaver, six-cylinder engines. The amenities of the *Luverne* included double-step running boards, Spanish leather seats, a gas-electric lighting system, a two-toned, deep brown paint job, sixteen-spoke wire wheels, and one of the first front bumpers in the industry. One of the auto's slogans was "Big and Long and Brown and Strong." Since the autos sold in the $3,000 range, another motto was "It is Worth the Money." *Luverne*s were generally heavy cars, built to withstand the rough country roads.

The *Luverne* was named after the city it was built in, Luverne, Minnesota. Al Leicher graduated from working at his father's carriage business in Loganville, Wisconsin. Relocating to Luverne at the age of eighteen in 1891, Al began working at his Uncle William "Fent" Kendall's carriage shop and purchased it from him in 1894. They continued making carriages until 1916, thirteen years after they also started manufacturing automobiles.

The *Luverne* advertising used a lot of testimonials from consumers who had purchased a *Luverne*. One of these was a farmer from Montana who stated that the *Luverne* was the perfect car for the mountains, inspiring the renaming of one of their models to the *Montana Special* with the slogan, "The Car for the Mountains." It had 50 horsepower and sold for $3,000.

The *Luverne* factory never employed more than twenty people and only built about fifty cars per year. The most famous, and most prized, was the *Big Brown,* which debuted in 1914 with a six-cylinder Beaver engine. In 1917, *Luverne*s switched to six-cylinder Continental engines. Most years, three models were offered: a surrey, a runabout, and a touring car.

The ad for the Montana Special

The 1912 Luverne Montana Special

The company began to branch out into building hearses, trucks, and fire engines. In 1912, they made their first truck. Autos were discontinued, and trucks dominated as the company name was changed to the **Luverne Motor Truck Company** in 1919. In 1922, as the company began to specialize in fire apparatus, the name was changed to the **Luverne Fire Apparatus Company**. As the fourth generation of Leichers built fire engines as the **Luverne Fire Equipment Company**, in 1985, the company was sold to the **Luverne Truck Equipment Company** of Brandon, South Dakota. This company was purchased in 1997 by the **Spartan Company** of Charlotte, Michigan.

A 1923 Luverne fire truck

The *Pan* automobile was made in St. Cloud, Minnesota, and was named after company owner and president Samuel Pandolfo. Pandolfo was a school superintendent at Las Cruces Schools in New Mexico. He started selling insurance on the side and realized how much more money he was making in that business than he did in school administration. He decided to enter the "private sector."

In 1916, he formed an auto company and incorporated it in Delaware. He started selling stock certificates for $10 apiece. He set up headquarters in St. Cloud, Minnesota, because it had access to two transcontinental railroads, was close to iron ore mines, and had a deep-water port in nearby Duluth, along with an abundant supply of electrical power. The **Pan Motor Car Company** was producing automobiles in 1916 that had high clearance because of the bad roads. The seats could be folded down into a bed, and the cars came equipped with a small refrigerator!

1919 Pan Model A

Touring cars and roadsters were manufactured with a four-cylinder engine. From 1918 to 1922, the **Pan Motor Car Company** produced 737 automobiles. The company fulfilled government contracts, including one where they developed a tank-tread tractor for the military.

Housing for the workers was built and called Pan-Town. The town included fifty-eight homes, a fire department, and hotel.

One of the remaining Pan-town (or Pantown) houses

Pan Motor Company Stock Certificate

Samuel Pandolfo was indicted for stock fraud in relation to his sales of company stock. He was cleared of those charges but convicted of mail fraud and served two and a half years of a three-year sentence. The company went under shortly after.

In 1903, Maurice Wolfe of Minneapolis received national press as the first salesman to sell a car, a *Cadillac,* to a Native American, Chief Big Mouth of the Billings, Montana, Crow tribe. Three years later, he joined with the Wilcox brothers, John F. and F. E., to form the **H.E. Wilcox Motor Car Company** and build a four-cylinder, double-chain-driven car with a sliding-gear transmission called the *Wolfe.* The cars were available with either an air-cooled or a water-cooled, 24-horsepower engine. From 1907 to 1909, the cars had the slogan "Absolutely Standard" and sold for $2,000.

In 1909, Maurice Wolfe relocated to Shelbyville, Indiana, and became involved with designing the autos the *Clark* and the *Meteor.* Subsequently, the name of the *Wolfe* auto was changed to the *Wilcox* in 1909. Its new slogan was "Built by Westerners for the West." The name of the firm, the **H.E. Wilcox Motor Car Company**, stayed the same.

The company sold 119 cars in 1909, and in 1910, it sold 189 with a tonneau model, which seated seven people, and a baby tonneau model, which seated five. The roadster model seated two.

The last year for the **H.E. Wilcox Motor Car Company** was 1910, when the company switched to making trucks exclusively. They continued making trucks until 1928.

The H.E. Wilcox Motor Car Company switched to trucks, or "trux," as they called them in ads.

The Wilcox Automobile

Chapter 24:

Going Green in Mississippi

GreenTech Automotive manufactured the electric car the *GTA* and was headquartered in Tunica County, Mississippi. The company was formed in 2003 and merged with Hong Kong-based **EuAuto Technology** in 2010. The company absorbed the **EuAuto Technology**'s auto, the *Mycar*. They then produced the *GTA Mycar*, which was a two-door, two-seat, sporty rear-wheel-drive electric car that lived up to its designation as a "Neighborhood Electric Vehicle," a special designation for battery-powered electric cars that go twenty-five miles per hour or less.

2014 GTA Mycar from Mississippi

GTA's cars were first built in Hattiesburg and later in Horn

Valley in Mississippi. The company merged with **VL Automotive** in 2014 and in September 2014 opened a new facility in Robinsonville, Tunica County, Mississippi, and began manufacturing **Fisker Karma**'s the *WM Destino*. This vehicle uses a solar panel to use the sun for fuel as well as using an electric, rechargeable battery.

Mississippi's GTA MyCar was short-lived

The GTA (GreenTech Automotive) MyCar factory closed in 2017.

In 1911, the **Southeastern Automobile and Machine Company** set out to build an auto in Hattiesburg, Mississippi. They were capitalized at about $50,000, but great doubt exists that they ever manufactured anything.

Chapter 25:

Diana, Dorris, and the Moon Motor Company of Missouri

THE **Moon Motor Company** added automobile manufacturing to the company's carriage business. The car was designed by Louis P. Mooers and was a four-cylinder, 35-horsepower model. In 1913, a six-cylinder auto was added, and the *Moon* autos changed to Continental engines. In 1919, the *Victory* was released with a radiator similar to the *Rolls Royce*, successfully appealing to consumers looking for a more luxurious car.

Actress Clara Bow on the hood of a Moon auto
and Babe Ruth and his Moon auto

Joseph Moon, the founder of the **Moon Motor Company,** started off making carriages with his brother John in 1880 as the **Moon Brothers Carriage Company**. Turmoil between the brothers caused them to go their separate ways. John started a separate carriage company.

Joseph Moon began making autos in 1905 and made cars and carriages simultaneously until about 1917, when the company began using the carriage-assembly space for building commercial vehicles. It was thought that building vehicles such as hearses, taxi-cabs, and aircraft would be more profitable than carriages. Unfortunately, they weren't.

The early *Moon* vehicles were designed by Louis P. Mooers, the well-respected designer of the *Peerless* auto. The early autos sported Mooers's overhead-valve, four-cylinder engines and sold well among the Midwestern farmers. This was due to the good reputation *Moon* had from building and selling carriages for twenty-five years.

One of the benefits of the *Moon* vehicles is that they stayed up to date on improvements, adopting improved ignition systems until becoming one of the first companies to start using advanced ignition systems. This included the Bosch high-tension magneto starter, the Disco acetylene starter, and, finally, the Delco system. They were also early adaptors in using aluminum and aluminum-coated steel in their cars as well as becoming early users of four-wheel hydraulic brakes and balloon tires. They were a leader in building a closed-body car in 1908 and in using their "Straight 8" eight-cylinder engines.

Moon Motor Car Company, St. Louis, Missouri

Moon trading card

Joseph Moon insisted on "quality over quantity" and customer service over profits, so he held annual car production down to around 2,500. The highest sales rank among car companies that **Moon Motor Company** ever rose to be was twenty-four, but **Moon Motor Company** had over seven hundred dealers and eventually sold autos in forty-seven countries, taking advantage of the fact that St. Louis was a major shipping center. In New York City, the *Moon* auto was sold by the **Hol-Tan Company** in 1908 as a *Hol-Tan*.

Joseph Moon died a wealthy man in 1919, and his son-in-law, Stewart McDonald, who had married Joseph's daughter Grace, took over the leadership of the company. He had previously been its vice president.

A 1925 subsidiary of the **Moon Motor Company** was the **Diana Motors Company**, with the same board of directors as the **Moon Motor Company**. The car, called the *Diana*, was built in St. Louis and came with an eight-cylinder Continental engine with bronze wire wheels as an option. *Diana*s ranged from $1,595 to $5,000.

The meticulousness that Joseph Moon brought to the company was gone in 1925, as the *Diana* was not properly road-tested before being released to the public. The *Diana*'s Continental engine had defective piston rings and an improperly placed water pump that caused overheating. The marque was discontinued in 1928 and replaced by the *Windsor* in 1929. This proved to be just another luxurious "Straight 8" vehicle produced by the **Windsor Corporation**, another "paper company" that unfortunately didn't garner the main company any profits.

Diana hood ornament and Windsor auto car badge

Both the *Diana* and the *Windsor* hood ornaments caused controversy. The *Diana's* controversy was due to the hood ornament's similarity to the *Minerva* of Antwerp, Belgium. The *Windsor* used a logo similar to the House of Windsor and was asked to change it, which it did.

When the *Windsor* was a poor seller, the **Moon Motor Company** was looking for something to do with their ample factory space at 4400 N. Main Street in St. Louis. One of the things they came upon was manufacturing manure spreaders, which they would have pursued further if not presented with an offer by Archie Andrews to build his car, the *Ruxton*.

Andrews manipulated the stock to get a controlling interest and fixed the board by adding members of his choosing and ousting the former board. Objecting to this, the longtime employees of the **Moon Motor Company** barricaded themselves inside the factory, locking the doors. A member of the St. Louis Police Department finally gained entry through a window, and the employees capitulated.

Andrews left the **Moon Motor Company's** business affairs in such a poor order that it wasn't until 1966 that a court finally wrapped it all up.

The **D.W. Haydock Automobile Manufacturing Company, Inc.,** was a St. Louis, Missouri, company that manufactured a car called the *Cosmopolitan* from 1907 to 1910. The car was a high-wheeler, powered by a one-cylinder, air-cooled engine and priced at $350. To facilitate hill climbing, it had a single chain added to the front axle, an early front-wheel drive. It was called the "Haydock Front Drive."

Another high-wheeler, similar to the *Cosmopolitan,* was the *Gleason,* manufactured by the **Kansas City Vehicle Company** in Kansas City, Missouri, from 1909 to 1914. The autos had two-cylinder, water-cooled engines. They didn't change their vehicle during their five years of production and kept the high-wheeler design, even though pneumatic tires were entering the marketplace. They sold out to the **Bauer Machine Company** in 1914, a company that wanted to make a cyclecar.

––––––––––––

In 1905, the **Auto Buggy Manufacturing Company** at 3993 Morgan Street, St. Louis, Missouri, manufactured the *A.B.C.*, a high-wheeler vehicle with friction transmission and a one-cylinder, 10-horsepower, air-cooled engine. It was advertised to be "as simple as A.B.C." The initials also stood for the engineer of the car, Amedee B. Cole. The company manufactured this vehicle through 1907.

In 1908, the company upgraded its vehicles by switching to two-cylinder, 18-horsepower, water-cooled engines. These autos were capable of going up to 30 miles per hour. They also changed their name to the A.B.C. Motor Vehicle Manufacturing Company. The car sold for $650 and was advertised as "the cheapest high-grade car in America."

––––––––––––

The president of the **St. Louis Car Company** (a railroad firm that manufactured train cars), Charles Kobusch, owned most of the stock in a new company he started, the **Kobusch Automobile Company**. In 1906, the company released *Model A*, *Model B*, and *Model C* of their two-cylinder vehicle, the *Kobusch*.

The *Model A* was a five-passenger touring model, also available as a demi-limousine. The *Model B* was a limousine model, while the *Model C* was a landaulet. The autos were meant to be similar to the French *Mors* car. The railroad company, the **St. Louis Car Company**, absorbed the company and bought the rights to produce the *American Mors* from the French company.

The **St. Louis Car Company** had also struck out on a car called the *St. Louis* in 1905. With the *American Mors*, the company was hoping that the third time was the charm. They did do well with the *American Mors,* producing it from the end of 1906 to 1909. They then switched to manufacturing their own vehicle again, the *Standard Six.* The *Standard Six* had a six-cylinder, 50-horsepower engine on a 124-inch wheelbase chassis. The three-speed car sold for $3,000 for each of its models: the touring car, the tonneau, and the roadster. For $4,000, the company sold a limo model.

In February 1910, the company moved its manufacturing from St. Louis to Wabash, Indiana, and changed their name to the **Standard Automobile Company of America**. Maybe they couldn't fit all that on their signs or pay for all those words in its advertising, but by 1911, they were gone.

The **St. Louis Motor Company** did business at 1211-13 North Vandeventer Avenue in St. Louis, Missouri. The company was founded by George Preston Dorris, who developed and patented the float carburetor, and John French in 1898. French oversaw marketing, and Dorris was the head of engineering and production.

1916 Dorris

A touring car at $1,500 and a runabout model selling for $1,200 were manufactured until 1905. The company moved to Peoria, Illinois, for its last few years and went defunct in 1907.

George Dorris left and started the **St. Louis Motor Carriage**

Company in 1905, at first marketing a car with a four-cylinder engine. In 1916, a six-cylinder engine replaced the four-cylinder engine. In 1906, the company name was changed to the **Dorris Motor Car Company**. The car, called the *Dorris* and designed by George Dorris, was considered luxurious, and the company made cars until 1923, although the autos were available because of left-over parts until 1926.

In 1907, in St. Louis, Missouri, the **Everybody's Motor Car Manufacturing Company** manufactured a car named the *Everybody's*. It was a 10-horsepower, two-cylinder runabout with a steering wheel, friction transmission, and a double-chain drive, selling for $450. The auto didn't change much in its two years of production except for the price, which went up to $500 shortly before the company went out of business. The auto wasn't for everybody, and soon, nobody had an *Everybody's*.

Another high-wheeler was manufactured by the **Victor Automobile Manufacturing Company** of 3936-3940 Laclede Avenue in St. Louis, Missouri. It was called the *Victor* and had a single-cylinder, water-cooled engine and a friction transmission. Advertised as "Hand Forged," they sold for $450. The company was in business from 1905 to 1911.

1907 Victor auto buggy

At 532 De Baliviere Avenue in St. Louis, Missouri, the **Success Automobile Manufacturing Company** manufactured high-wheeler autos and light delivery vehicles in 1906. Starting at one of the lowest prices to ever be charged for an auto, $250 (cheaper than the *Model T* at its cheapest, $260), the later models got more expensive as they became more complex. The car was available with a two- or four-cylinder engine and either air-cooled or water-cooled.

The company was founded by John C. Higdon, who built his first auto in 1896. The 1906 models sold to the public had steel tires, with rubber available for $25 more, and a two-speed planetary transmission.

The **Success Automobile Manufacturing Company** came out with *Models A* through *F*, each one a bigger horsepower than the previous one. Their success lasted until 1909, when the company went defunct.

Success ad

Russell Gardner founded the **Gardner Motor Company, Inc.,** in St. Louis, Missouri, in 1919. He made the "Banner Buggy" before turning to autos. His first auto had a four-cylinder Lycoming engine. In 1922, the **Gardner Motor Company** became the first and only car company to offer a one-year guarantee. In 1924, six- and eight-cylinder engines were offered. From 1926 to 1929, only straight eight-cylinder engines were offered for the *Gardner.*

The last *Gardner*s came out in 1930 and 1931 when a six-cylinder, low-hung, front-wheel-drive vehicle hit the streets in the swan song for the *Gardner* marque.

The 1930 Gardner eight-cylinder auto

One-year warranty issued by Gardner; the 1930 Gardner Sport Phaeton

Chapter 26:

The Montana Company of Reconditioned Cars

Feris J. DeVault operated the **DeVault Motor Company** at 109 S. B Street in Livingston, Montana. He produced cars in the 1930s that were "reconditioned" cars. DeVault, who called his cars *DeVaults*, produced an unknown number of original vehicles and sold them to local Livingston consumers. The cars were said to be stylish looking, resembling a cross between a *Ford Model A* and a *Plymouth*.

Although car manufacturers have been few and far between in Montana, the state still loves its trucks and racing cars.

Chapter 27:

Truck and Tractor Companies of Nebraska

Is it any surprise that a vehicle built in Nebraska was a truck or farm machine? The *Patriot* was one of the vehicles built on the plains because farmers needed the help of machines. The farmers might have owned a car, but they still did their plowing using farm animals to pull the plow.

The enterprising A. G. Hebb started the **A.G. Hebb Motor Company** in 1913, changing the name to the **Patriot Motors Company** in 1920. Hebb went through many jobs before becoming a truck manufacturer. In the early 1900s, he became an insurance agent with the Lincoln Insurance Company, selling hog insurance to farmers to guard against the hog catching cholera. Unbelievably ditching the hog cholera insurance business, upon marrying Daisy Jones, a farmer's daughter, he started a picture-framing business with her in Lincoln, Nebraska.

Hebb went on to manufacture wallpaper paste in the family firm before moving on to selling animal seed. After a few good years of seed sales, he noticed that less seed was being purchased year after year. He attributed this to fewer people using horses and farm animals to do chores and more farmers using autos and tractors instead.

Switching gears, he sold his seed business and started a successful **Ford** dealership. Besides **Ford** cars, he sold chase trucks. He then began building bodies for utility vehicles not being made by the larger auto companies. This included fire trucks, ambulances, and farm equipment. In 1913, the **Hebb Motor Company** employed ten carpenters constructing the hard, pine wood auto bodies. They expanded to putting out their own complete vehicles under the *Patriot* name. The vehicles carried patriotic names: Washington, Lincoln, and, later, Revere. From 1918 to 1919, they advertised as "Patriot FARM Trucks" with the word "farm" emphasized, symbolizing who Hebb figured the company's market was.

The **Hebb Motor Company** pioneered the use of a hydraulic gear to make the first dump trucks able to easily expel their cargo. They also manufactured buses, fire trucks, and custom-made vehicles. Like many Midwest car companies' products, the *Patriot* line was known for its toughness, able to navigate mud and ice.

By 1920, the auto company suffered under a sour economy and went into receivership. The company had a second act under the Woods brothers, who purchased the company and kept it going until 1948. A form of the company survives today manufacturing fire engines.

From the **Colby Motor Car Company** of Mason City, Iowa, came David W. Henry, who in 1912 convinced the politicos of Omaha, Nebraska, that they should finance a car that he had designed. It was a four-cylinder, 30-horsepower, touring car selling for $1,250. Calling the car the *Omaha* and the company the **Omaha Motor Car Company**, the cars delivered in 1911.

1912 Colby Underslung

After trying unsuccessfully to merge with several Iowa and Minnesota firms, it was said that the last year of the company was financed by a little, old lady who had inherited about $400,000 and was in the beginning stages of dementia. Whenever the factory needed money, someone would go to her and get a check to tide them over, but finally, by the end of 1914, the company was in receivership. Total production was about 5,551 autos over three years.

The **Drummond Motor Car Company** was founded in Omaha in 1906 as the **Drummond Carriage Company of Omaha**. In 1909, they decided to sell out the rest of their carriages and get into the auto-repair business. By 1916, they figured that if they could repair cars, they could build them also.

The first (1916) *Drummond Americans* were offered with either four cylinders or six cylinders in town car, runabout, or town car sedan, which had a detachable top. The 1917 models were all equipped with V-8 Herschel-Spillman engines. Their slogan was, "A Car In Which Every Part Is a Performer."

Logo for the Drummond Motor Car Company of Omaha

The 1916 Drummond American

The car may have performed well on the street, but in the marketplace, not so well. With 1916 production of 123 units and 1917
of 177, the company decided to change, starting with their name.
The new name was the **Douglas Motor Company** (for Omaha's
county), and it was said that the reason for the name change was to
avoid paying royalties to Drummond.

The new company took the same car, renamed it to the *Douglas,* upped the horsepower to 75, and increased the size to a 122-
inch wheelbase for 1918 and 1919. They had the sporty *Speedster
Special,* which had a 126-inch wheelbase, but by 1919, but it wasn't
enough to save the auto. The production for 1918 was 213 units,
but for 1919, it was down to 121. However, the company survived
making trucks, which they did into the mid-1930s.

CHAPTER 28:
NEVADA'S DESERT FLYER
COULDN'T

THE AUTO HISTORY OF NEVADA is surprisingly short. Since it takes a gambler to start an auto company and Nevada is full of gamblers, one would think that more Nevada car companies would exist! In 1907, the **Nevada Motor Car Company** announced that they would be building a factory in Reno, Nevada. George Bertzchi had invented the *Desert Flyer,* a four-cylinder car. It was mostly adapted from *Pope-Toledo* parts with a larger radiator and fuel capacity.

When George Bertzchi thought the company was being mismanaged, he quit, sued for back pay, and started the **Bertzchi Motor Company** in Council Bluffs, Iowa. He still failed to mass produce any autos. They at least produced a prototype. In 1897, Reno's W. G. Caffrey proposed the *Caffrey Electric,* an electric vehicle. The drawback was that it required overhead electric wires wherever it went!

Nevada is more known for racetracks, with the world's longest racetrack, than they are for auto companies.

Chapter 29:
The Hoyt Steamer of New Hampshire

NEW HAMPSHIRE HAD THE 1900 *Hoyt Steamer* automobile as its main claim to fame in the annals of auto history. Dr. Adrian Hoyt, a physician and Dartmouth graduate, and the **Concord Motor Coach Company** of Penacook, New Hampshire (near Concord), assembled the 1898 *Concord Steamer*, and a prototype was built in 1899 for the *Hoyt Steamer*, which came out in 1900.

1901 Hoyt Steamer

Dr. Hoyt at the tiller of one of his cars

Dr. Hoyt announced that he would soon be employing 150 to 250 men in an auto factory and received a ten-year local tax exemption from Concord officials. He claimed they would be assembling three autos per day, but this never materialized. The company made five steam cars and six electrics before making the determination that marketing electrical supplies was more profitable than automobiles. They became the Hoyt Electric Instrument Works and, as of 2019, are still in business.

CHAPTER 30:

THE MERCER AND MORE IN NEW JERSEY

EW JERSEY HAD SOME MEMORABLE marques, especially in the early days of the American auto industry. The *Riker Electric Car* was one of the first, winning one of the first auto races in the United States. The race was held in Narragansett Park in Providence, Rhode Island, in 1896, and the prize was $900.

Riker Electric Vehicle

Production was in Elizabeth, New Jersey, and not many were produced until 1901, when **Riker** merged with the **Electric Vehicle Company**, which had sold over two thousand vehicles in 1896, partly through providing Manhattan, New York, with electric cabs. The **Riker Electric Vehicle Company** was acquired by the **Pope/Columbia** consortium in 1902 and continued manufacturing electric vehicles until bankruptcy in 1907.

The *Mercer* was an automobile that was controlled by the Roebling family, Ferdinand Roebling, son of John A. Roebling (the engineer of the Brooklyn Bridge), and nephew Washington Roebling. Other stockholders were the Kuser family, who made their money in banking, bottling, and breweries. Starting in a vacant brewery in Hamilton, New Jersey, in 1909, the company moved to the state capital: Trenton, New Jersey. The car was named the *Mercer* after its location in Mercer County, New Jersey.

The **Mercer Automobile Company** was known for its race cars, starting with the 1910 *35R Raceabout*. This was a stripped-down, two-seat speedster with a curved dashboard, which gave it a streamlined look, and it could easily reach ninety miles per hour. The car was the victor in many races in the early 1900s and the rival to the *Stutz Bearcat* in racing.

1912 Mercer 35R Raceabout and the 1916 Mercer 22/72 touring car

1922 Mercer

Some of the race car drivers that drove *Mercer*s included Barney Oldfield and Ralph DePalma. The *Mercer* was considered a luxury car, and no more than one thousand were built a year. One of the last Roebling brothers, Washington A. Roebling, died on the *Titanic* in 1912, and this Roebling demise meant the company was sold to a Wall Street firm in 1919. They also purchased the old *Locomobile* and *Crane-Simplex* marques and became known for producing luxury cars, but they overextended and went into receivership in 1925. A couple of attempts to revive the marque only resulted in a few more being produced in 1928.

The **Simplex Automobile Company, Inc.,** company started in New York City, New York, as **S & M Simplex** in 1904 and went out of business in 1907. It was then reorganized as the **Simplex Automobile Company** in New Brunswick, New Jersey, in 1907 and produced cars for ten years. The most well-known **Simplex** model was the high-quality, chain-driven auto of 50 horsepower. By 1914, it was said to be the last chain-driven auto.

In 1915, under new designer Henry Crane, the style of the car changed into a less sporty model, one that was manufactured until 1917. The body of the *Simplex* was still being assembled onto a *Crane* chassis as late as 1921. In 1922, the company was reorganized as the **Crane-Simplex Company of New York** when the company combined with the **Crane Motor Company**. The new company produced autos for another two years before going defunct.

S. & M. "SIMPLEX."

The Simplex from New Jersey

The *Conover* was a four-cylinder, 40-horsepower vehicle built in Paterson, New Jersey, from 1906 to 1912. The **Conover Motor Car Company** produced this large car, available as a runabout or a five-seat tonneau. The secret identity of the **Conover Motor Car Company** was the **Watson Machine Company**, makers of wire rope machinery.

It was estimated that twenty-five of the cars were made, with the company losing money on each one, although they didn't officially close their auto department until 1912. The **Watson Machine Company** is still in the same place—44 Railroad Road in Paterson, New Jersey—still making wire rope machinery.

———————————

The **Automobile Company of America** in Marion, New Jersey, manufactured the *Gasmobile* from 1900 to 1902. It was a high-powered, high-speed carriage with a 35-horsepower, six-cylinder engine, which was actually two three-cylinder engines, deemed "futuristic" by the media, but it wasn't futuristic enough to prevent the car company from going out of business in 1902.

The 1900 Gasmobile

1900 Gasmobile Runabout

F. C. and Edwin Vandewater moved in 1909 to Elizabeth from Iselin, New Jersey, to continue manufacturing the *Eagle,* a car they had assembled and sold on a limited basis prior to 1909. Keeping the name **Vandewater and Company**, they decided to change the name of the vehicle to *Correja* for a probable "silent" investor.

The Correja; the Correja advertised on a billboard

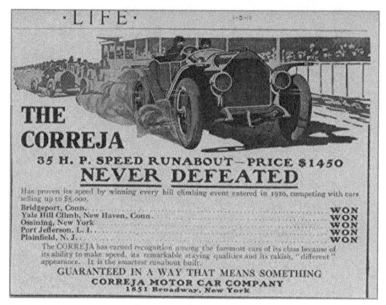

Correja advertisement

The *Correja* was introduced in 1909 and had a four-cylinder engine with shaft drive and 40 horsepower. It was heavy and well built, had a 110-inch wheelbase, and was available in a touring model or a runabout. Its motto was "Takes Every Hill, and Always Will." In 1910, torpedo models were available in four- to seven-passenger models, with 116- or 124-inch wheelbases and 50 horsepower.

For 1911, the only model released was a smaller, four-cylinder, two-passenger "speed runabout" with 35 horsepower and a 105-inch wheelbase.

In 1912, a six-cylinder torpedo model with a 125-inch wheelbase was added to the four-cylinder cars the company offered. The next year, 1913, they no longer offered the four-cylinder models and instead offered six-cylinder models with 65 horsepower on a 125-inch wheelbase.

The company's last year was 1914, when they declared bankruptcy. The only models released that year were a five-passenger touring car and a six-passenger *Knickerbocker* touring car.

CHAPTER 31:

THE FICTITIOUS CANNON BALL OF NEW MEXICO

ROUND 1919, IN TEXACO, NEW Mexico, a small town (population about 1,100) on the New Mexico–Texas border, the **Cannon Ball Motor Company** was formed by C. A. Roberson to manufacture the *Cannon Ball* (a.k.a. the *Cannonball*) automobile.

Before the production began, money was raised in the amount of $10 million—a large amount for a small town. Plans were bandied about regarding building a million-dollar factory in Santa Fe, New Mexico. The 1921 *Cannon Ball* was advertised but never seen.

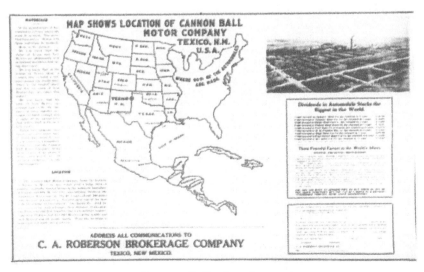

Cannon Ball advertisement

After appearing in a number of auto magazines and publications, the auto, and the men who raised the capital, disappeared in the dead of night with the money. Later, in 1921, a legal inquiry came up blank on where the money went. The New Mexico deception was probably the most successful of all the different flimflam scams around the United States. This example just illustrates how many con men were attempting to tap into men's desires to become associated with an automobile company.

Chapter 32:
Franklin, Mobile, and More New York Car Companies

THE EASTERN SEABOARD HAS HAD its share of auto companies, with Massachusetts and New York playing major parts in American auto evolution.

1899 Mobile

J. B. Walker and A. Lorenzo Barber bought the steam car patents from the Stanley twins in 1899. They could not agree on forming a car company together, so Barber went on to form the **Locomobile Company of America**, and J. B. Walker started the **Mobile Car Company**. Both cars were at first almost identical to

each other. **Mobile** offered twenty assorted styles in 1902 and fifteen in 1903. Over six thousand *Mobile*s were manufactured in all.

1909 Thomas Flyer

1907 Thomas Flyer

Edwin Ross Thomas got his start in 1896, selling a gasoline engine kit that could be used to motorize a bicycle. He graduated to selling the complete motorized bicycle as the *Thomas Auto-Bi,* considered the first manufactured motorized bicycle in the United States.

In 1902, Thomas decided to get into the automobile business. The **E.R. Thomas Motor Company** took over the **Buffalo Automobile and Auto-Bi Company** and renamed the car they had been selling as the *Buffalo* the *Thomas.* The 1902 *Thomas Model 17* had one cylinder, 8 horsepower, and two gears. The 1903 *Thomas Model 18* was available as a runabout.

In 1904, the *Thomas Flyer* was released. It had three cylinders and a planetary transmission. The car gained fame when a 1907 *Thomas Flyer Model 35,* with four cylinders and 60 horsepower, placed first in the 1908 New York to Paris race. It also became the first car to cross the United States in the process, taking forty-one days to do it.

In 1912, the company went into receivership and was purchased by the **Empire Smelting Company** of Depew, New York. They continued to make *Thomas Flyer*s by special order until 1918.

The Maltby Automobile Company of Brooklyn, New York, was organized in 1900 by B. F. Maltby and produced a light, two-seater auto, the *Maltby,* with a tiller for steering. The suspension was fully elliptical all around, and the front spring was transversely mounted. The engine was under the seat. The car ceased production in 1902.

The B.F. Maltby Garage in the early 1900s.

Henry Lozier had a very successful bicycle business with the **Lozier Manufacturing Company** building the *Cleveland* bicycle in his four bicycle factories and selling them in his Toledo, Ohio, store. He decided to get into the auto business in 1898 and sold his bicycle business to the **American Bicycle Company** for $4 million.

This certainly gave him enough money to get started, and he started researching cars all up and down the East Coast as well as studying European autos.

Henry Lozier Sr. died in 1903, but his son, Harry Lozier Jr., took up the search and manufactured first a motorized trike and then in 1905 released the *Lozier,* a car designed by John Perrin, based on the European styles of the day, most notably the *Mercedes.* From studying the finest European auto parts, Lozier was able to have Perrin construct an auto that, at least compared to the other autos around in 1905, was indestructible, even though the first bodies were solid aluminum. Landaulet and limo styles were predominant. The *Lozier* won the Elgin Road Race, the Vanderbilt Cup, came in second in the Indy 500, and won many other honors.

Because the Plattsburgh, New York, plant was limited, so was the yearly production of *Loziers*, and supply never met demand. Its largest production was six hundred in 1912. Some Detroit entrepreneurs convinced Harry Lozier to move the plant to Mack Avenue in Detroit in 1910 but he regretted the move when he was eased out of the company in 1912. He later came back with the *H.A.L.* (his initials) auto and the **HAL Motor Company**.

Taking over was Harry Jewett, also president of the **Paige-Detroit Motor Company**. Bowing out of the company at this time was Fred C. Chandler, who had been with the company since their bicycle days. He went on to form the **Chandler Motor Company**.

A less expensive car, the *Lozier Light Six,* sold for $3,500. A few other *Loziers* were attempted, including one with four cylinders, but by 1915, the company was in receivership. A few more attempts to keep the marque alive were attempted, including selling the Plattsburgh and Detroit factories and moving to a smaller factory in Detroit. The last year for new *Loziers* was 1918.

1910 Lozier

The **Mora Motor Car Company** built their *Mora* in Newark, New York, from 1906 to 1911. Their ads said they had the "World's Record Bonnet Sealed Hero." Their motto was "*Mora* Makes Good." A public favorite was the four-cylinder ractype, of which it was announced that "only one hundred will be made each year." (They probably never made more than one hundred of any model.) In 1908, a six-cylinder model was introduced. The *Mora*s were available in touring, limo, roadster, and ractype styles. The ractype seated three.

The *Mora* chief engineer William Birdsall invented the *BrownieKar*, a miniature car for juveniles. It was built by **Mora** under the anagram **Omar**.

However, near the end of 1910, trouble loomed as creditors demanded payment. They were forced into liquidation, and the factory and fifty assembled cars were sold at auction in 1911. The founder, Charles Mora, moved to Cleveland and went into commercial vehicle production.

The automobile named the *Only* had only one cylinder. It also had a ball-bearing-mounted crankshaft with a flywheel at each end. It was claimed it could go sixty miles per hour and would get thirty miles to the gallon, all for $700!

Incorporated in 1909, the **Only Motor Car Company** of Port Jefferson, New York, released only one-cylinder vehicles until 1912 and 1913, when they manufactured a four-cylinder vehicle. The raceabout version, called the *Metropole,* was the most popular. The company only made it to 1913 before they went into receivership.

H. A. Moyer started as a wagon builder in Cicero, Illinois, in 1875 and then moved his business to Syracuse, New York. In 1908, he designed his first experimental car. In 1911, he began manufacturing the *Moyer*, large, $2,000 to $3,000 vehicles, of which full-pressure lubrication was a feature. Both four-cylinder and six-cylinder cars were assembled. The motto of the **H.A. Moyer Automobile Company** was "All Roads Are Level To a *Moyer*."

The last year the *Moyer* was released was 1915 and then the money ran out. The company was never incorporated and was mostly a one-man operation, although over four hundred cars were built. The advertising was all written in first person!

Factory of the H.A. Moyer Automobile Company
in Syracuse; a 1911 Moyer

The **Hewitt Motor Company** had their headquarters in New York City on East 35th Street, just off Fifth Avenue. They purchased the Selden patent from the **Standard Motor Car Company** of New Jersey in 1905 when that company decided to stop manufacturing autos.

By 1906, the company, headed by Fred Hewitt, brought out their lineup: one-cylinder cars which featured limo, town car, and "little touring car" models that seated two. A four-cylinder car was offered in touring and limo models. The cars were called *Hewitts*.

In 1907, the company released what Hewitt claimed was the first V-8 engine on the market, although he probably shared the honor with the *Buffum Runabout* of Abington, Massachusetts, which claimed a V-8 engine in 1905.

When the company was absorbed by the **Metzger Motor Car Company** in late 1907, the Hewitt division thereafter concentrated on trucks in affiliation with **Mack Trucks**. Fred Hewitt in the 1940s was an engineering consultant for **Mack Trucks**.

The *Vaughn* was a sporty, four-cylinder, 30-horsepower auto named after race car driver Guy Vaughn, built by **Vaughn Car Company, Inc.** Vaughn usually raced for the **F.B. Stearns Company** and popularized the toy tonneau model for them. He decided he should have his own car and had it manufactured in Kingston, New York.

From 1910 to 1913, the car remained essentially the same. Releasing a six-cylinder, 128-inch-wheelbase, 35-horsepower car in 1914 didn't save the company. The principals, including Fred Moskovitz, who would go on to design the *Stutz Bearcat*, all left to pursue other auto interests. The Kingston factory was sold to the **Remington Motor Company**.

Most auto entrepreneurs get but one opportunity to develop a winning car. Although "third time's the charm" worked for Henry Ford, it did not work for three-timer William Irving Twombly. His first company in 1904 was the **Twombly Motor Car Company**.

All of Twombly's efforts were based in New York City, with a factory at 52 Broadway in Manhattan. Twombly's first auto was a four-cylinder, 28-horsepower steam car selling for $2,500 for the touring model and $3,000 for a limo model. The controls were all on the steering wheel, and the car could go up to fifty miles per hour (fifty-five miles per hour with a good tailwind). When the car was unable to make a profit, Twombly quit the company and moved to 12th and 51st Streets in Manhattan as the **Twombly Power Corporation**. In 1905, he developed a gasoline-powered vehicle but didn't manufacture it.

In 1910, he came up with the *Twombly "Quick Detachable" Car*. It was a large, two-cylinder steam car with easily detachable parts, including the power source, and easily converted to alternate models such as adding a roof to the touring model to create the limo version. Twombly envisioned not only selling a passenger car to the people but also thought the business had great possibilities to build taxicabs, livery wagons, and delivery vehicles.

The Twombly "Quick Detachable" Auto, which had detachable pieces and could become different models

When this vehicle proved unsuccessful, in 1913, William Twombly started what became his most successful endeavor, manufacturing a two-seater cyclecar that sold for $350. In 1914, the engine was increased to four cylinders, and the price was raised to $395. In 1915, the cyclecar was converted to a light car with a three-speed sliding transmission and was offered as a taxicab, a two-seat runabout passenger car, and a four-person touring car.

Moving production to Avondale and Nutley, New Jersey, it looked like the business was doing well until Twombly was sued into bankruptcy by an impatient stockholder, Reverend Stuart Dodge, in 1915.

In 1865, **Heinz, Pierce, and Munschauer** were making birdcages, iceboxes, and other household items in Buffalo, New York. Pierce bought out Heinz and Munschauer and renamed the company the **George N. Pierce Company** in 1872. In 1896, the company started making bicycles.

In 1900, an attempt at building a feasible steam car failed, but the attempt in 1901 to build a gasoline-powered auto resulted in the 2-horsepower *Pierce Motorette*. In 1902, the horsepower was increased to 3 horsepower, and in 1903, the *Arrow* appeared and had a 15-horsepower, two-cylinder engine.

Called the *Great Arrow* in 1904, it boasted 28 horsepower, and its power, as well as auto size, steadily increased over the years.

Pierce-Arrow

1932 Pierce-Arrow

1934 Pierce-Arrow

In 1909, the first *Pierce-Arrow* appeared and by 1914 was available in three varied sizes for the **Pierce-Arrow Motor Car Company**. The *Model 80* in 1923 was the first with four-wheel brakes. In 1923, over twelve thousand cars had been manufactured, and in 1929, **Pierce-Arrow** had their best year with 9,700 straight-8 cars manufactured, but in 1932, production fell to 2,692 cars, and in 1933, the handwriting was on the wall as a consortium of Buffalo businesses purchased the marque and remaining parts and continued to assemble *Pierce-Arrow*s until 1938.

Franklin Light Roadster

The first Franklin auto of 1902

One of the most well-known New York car companies was the **H.H. Franklin Manufacturing Company**. In 1917, the name was shortened to **The Franklin Motor Company**. Founded by Herbert H. Franklin in Syracuse, New York, it was a successful company producing a luxury car from 1902 to 1934, when the Great Depression caused them to fail. Many of the luxury car companies fell during the Depression years of the1930s. During their thirty-plus years manufacturing autos, **The Franklin Motor Company** produced over 150,000 cars, averaging eight thousand to fourteen thousand per year.

H. H. Franklin, with engineer John Wilkinson, developed an air-cooled engine in 1891. This led to building an entire auto. Later, Franklin handled the business, and Wilkinson was the chief engineer. The *Franklin* was the most successful car with an air-cooled engine until the *Volkswagen* came out. To prove the reliability of the air-cooled engine to not overheat, a "stunt drive" was completed from Walla Walla, Washington, to San Francisco.

1905 and 1912 Franklins

H. H. Franklin started as a newspaper publisher and then bought a patent for die casting, opening the **H.H. Franklin Manufacturing Company**, the first die-casting enterprise in the world. With the profits, he went into the auto-manufacturing business. Besides the air-cooled engines, *Franklins* revolutionized the automatic spark advance, the use of aluminum in car frames, and both six- and eight-cylinder engines. By 1915, some *Franklins* could get thirty-two miles per gallon and go sixty-five miles an hour.

The 1918 Franklin touring car and the Franklin Motor Company factory

By 1914, all four-cylinder *Franklins* were phased out, and only six or eight cylinders were offered. One person who preferred *Franklins* was aviator Charles Lindbergh. After the company was liquidated, the air-cooled patent went on to be used in *Franklin* aircraft engines. A six-cylinder *Franklin* helicopter engine was converted to water cooling and used as the engine by Preston Tucker in the *1947 Tucker* automobile. The *Franklin* assets were bought by Ward Canaday, who also had purchased **Willys-Overland**. He used the money to bolster the **Willys-Overland Company** and ultimately had remarkable success when they developed the first *Jeep* for the military.

The *Hatfield* was manufactured by the **Cortland Car & Carriage Company** in Sydney, New York. Lasting from 1917 to 1924, the *Hatfield* was a four-cylinder, gasoline-powered auto. Later models were powered by four- or six-cylinder Herschel-Spillman engines, with a selection of sport bodies available. They had two models: a touring car and a two-seater. Wire wheels and individual door steps were options.

Harry Palmer was from a wealthy family who got rich making barrels. He joined forces with Charles Singer, of the **Singer Sewing Machine Company**, to form the **Palmer & Singer Manufacturing Company** in Long Island City, New York, in 1908. Renting a showroom at 167 Broadway, New York City, they started as dealers for the *Simplex, Matheson,* and *Isotta-Fraschetti* before switching to producing their own autos in 1908.

Using the former *Matheson* plant in Forty-Fort, near Wilkes-Barre, Pennsylvania, the new car was at first called the *Palmer-Singer,* or the *P & S,* later shortened to the *Singer* in 1909. The first offering from the new company was the *Skimabout,* a four-cylinder, 28-horsepower vehicle. It was followed by a six-cylinder, 60-horsepower car, available in toy tonneau and later in touring, limo, landaulet, and runabout models.

Henry Palmer died in 1911, but the company stayed alive for a few more years and a few more models. In May 1913, they struck up a deal with the **Maritime Motor Company** of St. Johns, New Brunswick, Canada, to manufacture the *Maritime-Singer Six,* an international car with components made in Long Island, New York, and assembled in St. Johns, New Brunswick.

In the company's last years until bankruptcy in 1914, the *Brighton* and *Magic Six* models were popular. The factory was briefly used to manufacture the *Owen-Magnetic* and then Charles Singer repurchased the factory at the end of 1914, replaced the engine in the *Magic Six* with a Herschel-Spillman six-cylinder engine, and renamed the car the *Singer.*

Now a car with 50 horsepower, guaranteed to do seventy-two miles per hour on the road, the *Singer* was built by the **Singer Motor Company** of New York, New York, in a plant at West 57th

Street in Manhattan. They were considered luxury cars, and prices could reach $9,000, depending on selected options.

In 1919, the **Singer Motor Company** moved to a larger plant in Mount Vernon, New York, for 1919 and 1920. In 1920, an auto with a V-12-cylinder engine was released, but it was a last gasp as the company went into receivership in November and declared bankruptcy in December 1920. In 1921, the last of the 1920 models were sold. Charles Singer died the year after, in 1922.

In 1908, the **De Schaum Company** manufactured a two-cylinder, 10-horsepower, friction-driven high-wheeler auto in Buffalo, New York, called the *De Schaum*. When it didn't sell, in 1909, they changed the name to *Seven Little Buffaloes*, but it didn't do any better—they only made fifty-four autos in all.

The **Buffalo Electric Carriage Company** made a wide range of styles of electric cars from 1901 to 1915. The cars could go seventy-one miles per charge and usually took forty-five minutes to fully charge. In 1906, the car company was purchased by Francis Babcock and renamed the **Babcock Electric Carriage Company**, with cars called the *Babcock*. In 1912, the company joined the **Buffalo Electric Vehicle Company**.

The **Buffalo Electric Vehicle Company** was a unification of car companies that combined in 1912. They included the **Babcock Electric Carriage Company**; the **Buffalo Automobile Station Company**; **Van Wagoner**, which was a truck company; the **Clark Motor Company**; and the **Buffalo Electric Carriage Company**. Francis Babcock became its first president, although the new company and cars reverted back to being called *Buffalo* instead of *Babcock* in the name.

By 1912, the **Buffalo Electric Vehicle Company** could see that gasoline cars were the wave of the future and advertised their vehicles as "designed along the lines of the latest gasoline cars." At first, the only style offered was the *Stanhope*, similar to a horse-drawn carriage (without the horse). In 1912, after a short sabbatical, the **Buffalo Electric Vehicle Company** came back with coupe, roadster, and brougham styles. The company manufactured electric cars until 1915, then continued manufacturing the trucks of **Van Wagoner**. The factory remains in Buffalo, repurposed as an art supply store.

The onetime factory of the Buffalo Electric Vehicle Company; Buffalo Stanhope advertisement

Another very early New York car company, founded in 1899, was the **Century Motor Vehicle Company** of Syracuse. They produced cars named the *Century*, a steam car with a 4-horsepower, two-cylinder vertical engine that used a bevel-gear drive instead of the more common single-chain drive. An earlier prototype car by the company was a single-cylinder, chain-drive, and tiller-steered model. The *Century* steam car was succeeded in 1903 by the *Century Tourist*, a gasoline-powered car, the last year the company made cars.

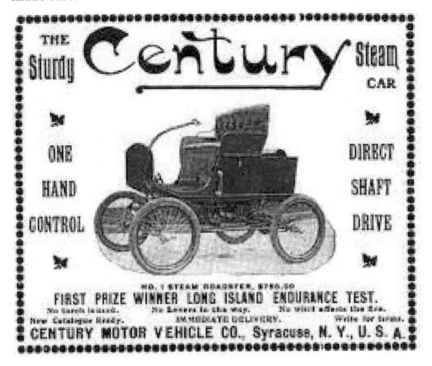

From 1901 to 1907 in Lockport, New York, Byron V. Covert and the **Covert Motor Vehicle Company** produced the *Covert*, a two-cylinder, chain-driven steam auto. Later engines offered were a 6-horsepower, single-cylinder engine and a four-cylinder, 24-horsepower model.

1903 Covert Chainless

In 1903, he introduced the *Covert Chainless*, a gasoline-pow-ered car with a sliding-gear transmission and the gearbox attached to the rear axle. In 1904, he dropped the two-cylinder engines and switched to four cylinders exclusively. Although they won the Buf-falo to St. Louis Endurance Run, the auto company went out of business in 1908 when Byron Covert decided to spend his time exclusively on developing and selling axles and running-gear com-ponents.

The **Playboy Motor Car Corporation** of Buffalo, New York, manufactured the *Playboy* auto from 1946 to 1951. This compact (ninety-inch wheelbase), three-seat convertible auto was driven by a four-cylinder, 40-horsepower Continental engine with an auto-matic transmission. It could, according to its publicity, go from 0–30 in 6 seconds and from 0–70 in 17 seconds. The brand was undercapitalized, and bankruptcy in 1951 ended the marque after only ninety-seven autos were manufactured.

The Playboy car, hood badge, and factory

An early steam vehicle was the *Lane* from Poughkeepsie, New York. The **Lane Motor Vehicle Company** formed in 1905 from a hardware specialties company and built five of their steam vehicles the first year. The *Lane* was another automotive brother act, built by siblings George, William, and John Lane.

The first models were smaller cars, but they soon arranged to build a larger vehicle. By 1909, their most successful year, they manufactured over two hundred autos with touring, roadster, and runabout models sold. The *Lanes* were mostly two-cylinder, 8- to 10-horsepower, chain-driven vehicles.

*This 1901 Lane Light Steam Runabout is
the only known remaining Lane vehicle*

Almost successful was the **Birmingham Motor Company** of Jamestown, New York, which manufactured the *Birmingham* from 1920 to 1923. The auto had a Continental six-cylinder, 55-horsepower engine with a swing-axle suspension. The company produced about fifty autos despite a stock swindle involving the company's president and mayor of Jamestown, Samuel Carlson, in 1923. Carlson was serving as president without salary.

1921 Birmingham

Much of the turmoil of the company involved stock sales. A stockholders' meeting of the company ended in a fight where a stock salesman was stabbed to death! Another company official smashed his way through a plate-glass window to escape the conflict. They subsequently were out of business by 1923 despite an attempt in 1924 to move the company to Canada and rename the auto the *Wright*.

Chapter 33:

North Carolina's Corbitt Auto Company Makes Trucks

T HE **CORBITT AUTO COMPANY** OF Henderson, North Carolina, produced an auto named the *Corbitt*. Operating from 1912 to 1913, the auto had a four-cylinder engine and was available in three body styles on a 120-inch wheelbase. The *Model A* was a two-seat roadster and cost $1,750, and the *Model C* was a five-passenger auto selling for $1,800. The name *Corbitt* lived on in trucks until 1958.

North Carolina has several firms that expressed a desire to build a car or even incorporated but never went beyond producing a prototype, sometimes not even that. Some of these planned cars were the *Asheville, Sawyer,* and *Grove Park* from Asheville, North Carolina; the *Wizard Junior* from Charlotte; the *Goodmobile* from Connally Springs; the *Durham* and *Lyon* from Durham; the *Armond* from Fayetteville; the *American Southern, Greensboro,* and *Vaughn* from Greensboro; the *Maxton* from Maxton; the *Piedmont* from Monroe; the *Waters* from New Bern; the *Alston* from Pittsboro; the *Brewer* from Raleigh; the *Waynick* from Reidsville; the *Leonard* from Salisbury; the *Cook* from Samson County; and the *H.E.* from Winston-Salem. None of these auto dreams from North Carolina, home of the first plane ride, ever came to fruition.

Corbitt vehicles in 1913

Chapter 34:

North Dakota's
the Holland and the Hammer

In 1901, the **Maddock Implement** Company was formed by Julius Hammer in Maddock, North Dakota. In 1904, Hammer introduced his 1904 vehicle with his patented "new idea in gas engines." This vehicle was basically an engine strapped to a farm wagon and some pulleys and belts. Hammer continued to experiment until around 1906.

Sam Holland would build cars for family and friends and didn't keep track of how many he had made. He started as a blacksmith in Park River, North Dakota. In 1898, his first car was a high-wheeler, steam-powered vehicle that could seat four. He then produced several gas-powered, single-cylinder tricycles. His next invention was a 6-horsepower, gas-powered, four-wheeler runabout that seated four, followed in 1905 by a four-cylinder, 12-horsepower, air-cooled engine vehicle called the *Holland*. He considered mass-marketing the *Holland* using the slogan "Patronize Home Industry. If You Don't, You Ought To." He stopped making vehicles in 1908.

In 1902, Frank Jaszkowiak built the first auto of Bismarck, North Dakota. It had a two-cylinder, two-stroke, 3-horsepower engine mounted to a two-seat motor buggy. Even though the car could only go eight to ten miles per hour, he still smashed into a tree on his test run. This was undoubtedly North Dakota's first auto accident. His next car was a two-seat motor buggy with 8 horse-power that could go twenty-five miles per hour. His third auto in 1908 had a two-cylinder, 20-horsepower engine with a battery to operate the electric headlamps.

Besides cars, Frank Jaszkowiak also invented a self-propelled well-digging rig, a cement and gravel well strainer, and a furnace, and he also produced and sold gasoline engines for farm threshing machines and ferry boats.

Frank Jaszkowiak's automobile, courtesy of the State Historical Society of North Dakota

CHAPTER 35:

CLEVELAND AND THE MANY OHIO AUTO MAKERS

ROM 1896 TO 1907, CLEVELAND turned out more cars than anywhere else in the United States. The first Motor City was Cleveland, which, along with northeastern Ohio, had not only begun to produce its own auto brands but, by 1907, was the nation's leading manufacturing center for auto parts, including pistons, valves, and frames, all products the area still produces.

In the first decade of the twentieth century, the car-manufacturing action had just moved to the Midwest as Cleveland, Ohio, was the city with more car companies and car-parts companies than anywhere else in North America. Cleveland remained a top auto-parts center and, with Indianapolis, Indiana, and finally Detroit, Michigan, became part the Auto Engineering Triangle, the centers of auto production. The state of Ohio had more than 115 car companies. The immediate surrounding states—Illinois, Minnesota, Wisconsin, and New York—also had a lot of auto production.

One of the leaders of the advent of the auto industry in Ohio was Alexander Winton, who started in his bicycle shop. He built gasoline autos and drove one of them from Cleveland to New York in 1897, generating good publicity for the motor car in general, which was still facing considerable competition from the horse and carriage. A 1903 Winton was the first to make a transcontinental tour; it went from New York to San Francisco.

Winton dissolved his company in 1924. He was the first to have "premade" cars for customers instead of having cars made to order. Winton also employed the first mass-production techniques in the auto field.

Winton hearse

1906 Winton

In Hamilton, Ohio, the **Imperial Motor Car Company** discovered the existence of another **Imperial** in Jackson, Michigan, and changed their name to the **Republic Motor Car Company** in 1910. They hadn't realized that another **Imperial** company existed because most of the investors were not from auto backgrounds: one was head of a safe company, another ran an ice-delivery service, and another built steam engines.

With the motto "Classiest of All," the car called the *Republic* had given itself high expectations. The car was built in an old paper factory purchased in 1909. A *Republic* had a four-cylinder, 40-horsepower, T-head engine with shaft drive, selective transmissions, and offset crankshafts.

A six-cylinder model was brought out in 1914 and was very popular, but because of World War I shortages, the marque didn't make it past 1917.

Lee A. Frayer and William J. Miller built the *Frayer-Miller* auto in conjunction with the **Oscar Lear Automobile Company** and the **Buckeye Motor Company** in Columbus, Ohio. The first cars had four-cylinder engines with a rotary blower said to keep the car cool in the hottest weather.

1908 Frayer-Miller "B" Touring Car

Frayer-Miller advertisement

On November 1, 1905, the company claimed that the first six-cylinder auto sold to a customer, William Monypeny, was a *Frayer-Miller*. Thereafter, the autos were available with both four- and six-cylinder engines in tonneau, roadster, and limo models. The car company name was changed to the **Buckeye National Motor Car Company**.

In 1907, the company sold trucks and, in 1908, an agricultural vehicle named the *Farmobile*. In 1909, they announced that they would be suspending production on their passenger cars in order to concentrate on commercial vehicles. In 1910, the factory was sold to the **Kelly Motor Truck Company**.

C.R. Patterson & Sons of Greenfield, Ohio, were the creators of the *Patterson-Greenfield* automobile. The company first started as a carriage maker and then began manufacturing their own motorized vehicles.

The company's organizer and namesake Charles Richard Patterson was born a slave in Virginia in 1833. His son and inheritor

Frederick was the first, and probably, to date, only African American to own an auto company.

C. R. Patterson offered over twenty-eight models of carriages when he died in 1910. Control of the company passed to Frederick, the first black to play football for Ohio State University. He was also the vice president of the National Negro Business League, founded by Booker T. Washington.

As autos began to come into vogue, carriage makers like Frederick Patterson could see the handwriting on the wall. Like very many carriage makers before them, the company decided to try its hand at manufacturing automobiles.

Building their autos from 1915 to 1918, the company first advertised a five-passenger vehicle selling for $685. The company expanded to manufacture coupes and sedan models of the *Patterson-Greenfield*. These came with a 30-horsepower Continental four-cylinder engine, electric starter and electric lights, a full floating rear axle, cantilever springs, and a split windshield, allowing for ventilation. After the company stopped auto production in 1918, the garage continued to make bus and car chassis.

Frederick Patterson with one of his first autos

The **Hoffman Automobile Manufacturing Company** manufactured a light steam car in Cleveland in 1903. It used a 6-horsepower, two-cylinder engine with a single-chain drive. They also made a gasoline car with a one-cylinder engine.

In 1904, the company changed their name to the **Royal Motor Car Company**. They manufactured a four-cylinder, 48-horsepower vehicle called the *Royal Tourist*. In 1911, the company combined with the **Croxton-Keeton Motor Car Company** and formed the **Consolidated Motor Car Company**.

H. A. Croxton (who built the *Jewel*) and F. M. Keeton combined the **Croxton Motor Company** of Massilon, Ohio, with the **Keeton Motor Company** to form the **Croxton-Keeton Motor Company** in 1909 to 1910.

They produced two models, the "German" and the "French." Business being fair, the two partners separated, and the company combined with the **Royal Motor Car Company** to form the **Consolidated Car Company**.

The **Consolidated Car Company** only lasted for a few months when in 1912 the struggling company was purchased by the **F.B. Stearns Company** of Cleveland.

The **F.B. Stearns Company** was founded by Frank Ballou Stearns (1879–1955), who put to use a workshop his wealthy (from stone quarries) father bought for him by building his first car in 1896 at the age of seventeen. In 1898, young Stearns combined with the Owen brothers, Raymond and Ralph, to manufacture autos. Their first auto, in 1900, had one cylinder and, like most cars of that time, was built like a buggy.

By 1902, the firm had sold about 150 of the cars and added two-cylinder, 20-horsepower touring models to the company's catalog. The Owen brothers left to pursue their own auto ventures (manufacturing the *Owen Magnetic*), and Frank Stearns continued the company. The 1902 model featured the first steering wheel on a car. Up to this point, autos had been steered by tillers. Curiously, the steering wheel was put on the left side, and the *Stearns* was the first in this aspect also, but in the next year, 1903, the steering wheel was placed on the right, not to be on the left again in *Stearns* autos until 1914.

Frank Stearns

The *Stearns* was increasing its horsepower each year and in 1904 introduced a four-cylinder car. In 1905, specialization to more of a luxury car began as only one, seven-passenger touring model was offered. It had a 118-inch wheelbase and a $4,150 price tag.

Stearns was also updating its transmissions each year and by 1906 had a four-speed gear shift. The car was considered a luxury car, although it did well in races and endurance contests, and its motto was "Runs Like a Deer." The *Stearns* prices ranged from $3,200 for a *Baby Stearns* to $7,500 for a four-cylinder *Stearns*.

In 1911, the **F.B. Stearns Company** agreed to the terms to use the first Knight engine. The invention of Charles Knight of Chicago, he charged $100 per car sold with his engine in it. The Knight engine used movable sleeve valves to cover the engine's intake and exhaust ports. The previous engines had used the poppet valve-train. The name of the *Stearns* was changed to *Stearns-Knight* to reflect the change to a Knight engine.

To build the new engine, the **Stearns-Knight Company** purchased the factory of the defunct **Royal Tourist Motor Company**. In 1913, a six-cylinder engine auto was added, and in 1914, the car was given a design update, with more European styling. Coupe and sedan models were added.

In 1915, an inexpensive *Stearns-Knight* was released, the *Little Four,* which sold for $1,750. In 1916, V-8s replaced the six-cylinder engines.

The 1929 Stearns-Knight Limo Model, the last year for the marque

In 1917, Frank Stearns reported that he was too ill to continue running the company, and, at the age of thirty-seven, resigned as president. The company was by this time assembling over four thousand cars per year. With new president George Brooker at the helm, not much changed except that the company changed back to four-cylinder models exclusively for three years. In 1924, a new six-cylinder car was introduced. After twenty years, the *Stearns-Knight* was still considered a high-end car brand (with a high-end price tag).

In December 1925, John North Willys of the **Willys-Overland Motor Company** purchased the **F.B. Stearns Company**. He set up H. J. Leonard as president, the man who was head of the company that manufactured the *Stephens* automobile. Willys promised he wouldn't be making any major changes to the *Stearns*. He wanted it to become the luxury marque for the **Willys-Overland Motor Company**. By this time, many different models of the *Stearns-Knight* were offered, including toy tonneau, limousine, and landaulet models.

For 1916, some of the wheelbases of the cars expanded to 145 inches, and only six- and eight-cylinder engines were available. Adding a car with an inline 8-engine and 100 horsepower for 1927, Willys hadn't realized that 1925 would be the last profitable year for the *Stearns* marque until the year-end 1926 total sales figures came in.

The company had begun losing money, and by 1929, after the beginning of the Great Depression, it didn't make sense to Willys to have a luxury, high-priced marque anymore. The stock had gone from $10 to 12½ cents, and ten days after, the board voted to dissolve the company, so the *Stearns* brand was discontinued and consigned to Cleveland auto history.

Frank Stearns overcame his ill health and lived to the age of seventy-six. He invented a two-stroke, overhead cam, diesel engine and sold it to the U.S. Navy. Before he died in 1955, he had started researching organic farming.

The **Schacht Manufacturing Company** of Cincinnati started making buggies in the early 1800s. By the turn of the century, they had started manufacturing auto parts. In 1904, they released their first auto, a two-cylinder, 10-horsepower runabout high-wheeler that reflected their start in buggy making. The engines were water-cooled and had steering wheels, not tillers.

In 1905, a four-cylinder car was added to the lineup. The *Schacht* was appreciated for its durability and was called "The Invincible Schacht." It was known for its brassy radiator in front. In 1909, the company name was changed to the **Schacht Motor**

Car Company.

"The Invincible Schacht" 1910 Model R runabout

After making over nine thousand autos in nine years, in 1915, the car line was discontinued in favor of trucks, which were produced by the **Schacht Motor Car Company** into the late 1930s.

———

Fred C. Chandler had worked for the **Lozier Motor Company** in Detroit and quit in 1914 to move to Cleveland and form his own company with a few others. This included C. M. Emise, who the company was originally going to be named after, but the **Chandler Motor Car Company** it was named, and the *Chandler* auto became a popular, medium-priced auto, even though it was very similar to the *Lozier*, especially the radiator.

The *Chandler Six* was the most popular of the *Chandlers*, and in 1915, a *Chandler* was driven from Tijuana, Mexico, to Vancouver, British Columbia, a trip of over two thousand miles, with no repairs. A *Chandler* won a contest for climbing Pike's Peak.

1919 CHANDLER driven by Dorothy Provine on "Roaring Twenties" on display at Movie World, Buena Park, California.

In 1919, the **Chandler Motor Company** started the **Cleveland Motor Company** to manufacture a smaller, more inexpensive car called the *Cleveland*. The brand went under in 1926, and the subsidiary was absorbed by the mother company, causing them to rename the company the **Chandler-Cleveland Motor Company**.

The peak year for *Chandler* sales was 1925, when sales topped twenty thousand units. In 1927, an eight-cylinder engine was introduced, but after a big loss for the year, the company announced it was being absorbed into the **Hupp Motor Car Company**.

The *Gaethmobile* and the *Gaeth* were inventions of Paul Gaeth of Cleveland. From 1905 until the **Gaeth Automobile Company**'s demise in 1911, they produced several different auto models.

Paul Gaeth put together his first car, a steamer, in 1898. His next car was a gasoline model. He worked at the **People's Automobile Company** in Cleveland from 1901 to 1902, when the company went bankrupt. Gaeth wasted no time in organizing the **Gaeth Automobile Company** to custom-make autos on demand for customers. From 1902 to 1904, about twenty-five of these one-cylinder, tiller-steering, chain-driven, runabout cars were made, called

Gaethmobiles. They were sturdy but small, with only a seventy-two-inch wheelbase.

In 1904, Gaeth began to make larger cars with two- and four-cylinder engines. The new vehicle name was shortened to *Gaeth*, and their motto was "Go With a Gaeth." Gaeth had offered runabouts, touring cars, and even a limo style when in 1910, his company was purchased by the **Stuyvesant Motor Car Company of Cleveland**.

Gaeth stayed active in the auto industry and produced some patents, including a fuel gauge. He made a small fortune in his later years restoring classic cars.

The *Great Eagle* of Columbus, Ohio, started off with an announcement in October 1909 by the **United States Carriage Company** that they would henceforth engage in the manufacture of hearses, ambulances, taxicabs, and passenger vehicles. Making its debut in 1910, the *Great Eagle* was a large vehicle, ranging from 135-inch to 142-inch wheelbases on their autos.

In 1912, a six-cylinder model was added to the four-cylinder cars offered in 1910 and 1911. The models available were touring, landaulet, and limo. Most of the cars could easily carry seven passengers.

The cars were well received, but in 1915, the company president's wife Katherine Myers (wife of Fred Myers) called a $6,000 loan and sent the company into a receivership, from which they never recovered.

The **Stuyvesant Motor Car Company of Cleveland** started off as the **Stiverson Motor Car Company**, named after founder Frank Stiverson, but he changed the name to the more commercial "Stuyvesant" in 1911. (He later changed his own last name to Stuyvesant.) Stiverson started as a sales manager for the *White*

Steam Auto and a manufacturer of nuts and bolts. When he decided to start auto manufacturing in 1910, he purchased the **Gaeth Motor Company**.

Stiverson then developed what he said was the first "en bloc" (cast as a single unit), six-cylinder engine. It was used in the first *Stuyvesant* car, the *Model 6-60*, a six-cylinder, 60-horsepower touring model auto. The company also continued to manufacture what had been the *Gaeth* as the *Stuyvesant Four* so that the company could offer both a six- and a four-cylinder engine in their vehicles.

By 1911, the plans to produce about three hundred autos went askance as the company began to have financial difficulties. They were first purchased by the **Grant-Lees Manufacturing Company of Cleveland.** In 1912, since this company was already manufacturing an auto called the *Grant Six,* it didn't want to produce another six-cylinder auto, so the *Stuyvesant Six* was shelved. The restyled *Gaeth,* the *Stuyvesant Four,* was still offered.

In 1913, the company was sold to the **Benton Manufacturing Company**, who restored the *Stuyvesant Six* to the marketplace while moving the company to Benton, Illinois. After that, the company disappeared.

The *Firestone-Columbus* was manufactured in 1909 by the **Columbus Buggy Company** of Columbus, Ohio. It was after their production of an electric car, the *Columbus Electric*, in 1903 and the *Columbus High-Wheeler* in 1907. The slogan of the high-wheeler was "A Car For the Masses, Not a Toy For the Classes."

The *Firestone-Columbus* was named after Clinton DeWitt Firestone, the **Columbus Buggy Company**'s president. The four-cylinder cars were available in many models besides the usual touring and runabout. Surrey, torpedo, and limo models were also available.

The company was trying to work through a bankruptcy at the end of 1913 when Clinton Firestone died of apoplexy, which effectively killed off the company for good. The buggy portion was sold and moved to Louisville, Kentucky, and the auto portion ended up in the hands of the **E.R. Thomas Company**.

The *Napoleon* was not directly named after Napoleon Bonaparte but for the town of Napoleon, Ohio. The 30-horsepower, four-cylinder car didn't change much from its beginning in 1916 to its Waterloo in 1919.

Although many people were proud that Napoleon had its own auto company, it wasn't enough to keep the **Napoleon Motor Manufacturing Company** in Napoleon. The company got a better offer from the good people of Traverse City, Michigan.

In 1916, Traverse City was about to lose one of its two largest businesses, the Oval Wood Dish Company. The city offered the **Napoleon Motor Manufacturing Company** the company's old factory space for free for three years, along with $75,000 in capitalization. In 1917, the **Napoleon Motor Manufacturing Company** jumped on the offer and changed their name to the **Traverse City Motor Car Company**.

For the rest of 1917, they manufactured the same car, still selling for $735. In 1918, they once again reorganized, this time as the **Napoleon Motors Company**, and added a six-cylinder auto to the lineup as well as a roadster model.

In 1919, the company was recapitalized for $500,000 and brought out their first truck. Due to its success, they announced that they would no longer be making passenger cars. They built about three hundred cars in total. The company continued to manufacture ¾-ton trucks until 1923.

The Napoleon

The **Jordan Motor Company** began in 1916 in Cleveland and operated until 1931, and their most productive year was 1926, when they manufactured and sold eleven thousand cars. Their plant was located east of downtown Cleveland at 1070 E. 152nd Street. This was a good location for them, right by the railroad. This provided easy access to shipping supplies in and autos out. In the first year for *Jordan*s, 1916, they sold over one thousand cars.

Ned Jordan

The company was founded by Edward "Ned" Jordan, who was a former advertising executive with the **Thomas B. Jeffery Company**. Most components of the *Jordans* were manufactured by other firms—the engines were by Continental, and they used Bosch ignition systems, Timken axles, and parts from other firms. The *Jordan* was a good example of an "assembled" car. It was known for its styling and multitude of colors. The red shades included "Apache Red," "Savage Red," and "Mercedes Red." Other colors included "Venetian Green," "Egyptian Bronze," "Liberty Blue," "Briarcliff Green," "Submarine Gray," "Chinese Blue," and "Ocean Sand Gray."

Jordan logo and a 1931 Jordan

The **Jordan Motor Company** was known for making limousines, which had six-cylinder engines until the 1931 *Model 90*, which had eight-cylinder, 85-horsepower engines. Some model names included the *Playboy*, the *Tomboy*, the *Sports Marine*, and the *Friendly Three*, which had the slogan "Seats Two—Three If They're Friendly."

The *Jordan* fell into the luxury category both by design and price. Models were priced around $5,000 on average, although some less expensive models were constructed. They fell victim to the diminishing market of people who could afford a "luxury" car and closed in 1931. Their last vehicle, the *Little Custom*, a luxury compact, didn't catch on with the public, accelerating the company's downfall. It isn't known exactly how many *Jordan*s were manufactured, but estimates range all the way from thirty thousand to one hundred thousand units sold.

Thomas White invented a sewing machine in 1858 and started manufacturing the "New England Sewing Machine" for $10. The business was so successful that White moved from Massachusetts to Cleveland, Ohio, to be closer to suppliers. When White incorporated the company in 1876, the name was changed to the White Sewing Machine Company.

In 1900, Thomas's son Rollin invented a steam car, which was a chain-driven, two-cylinder auto that used a tiller. Going into production in 1901, 193 were sold the first year. In 1903, a steering wheel was introduced to the car, along with shaft drive. The cars could run a hundred miles before the water had to be filled.

The original sewing machine company continued to diversify, making roller skates and kerosene lamps, but the sons of Thomas White were more interested in the autos and spun off the **White Motor Corporation** from the White Sewing Machine Company.

A 1910 White Steam touring car

The auto subsidiary was very successful, and *White Steam Cars* became well known and popular. The first car in the White House was the *White Steamer* that Theodore Roosevelt used. President William Howard Taft had a 1909 *White Model M Steamer* added to the White House car pool.

A 1909 White Steam Car, the first car in the White House fleet

In 1908, two models were offered, the 20-horsepower *Model L* and the large, seven-seater *Model K*, which had a wheelbase of 122 inches. White was producing over one thousand cars a year, with 1906 being the largest production year with 1,534 cars sold.

In 1910, gasoline vehicles began to be sold along with the steamers. Electric lighting and six-cylinder engines were offered with the cars, which were produced until 1918, by which time the **White Motor Company** had become firmly established as a truck manufacturer. They continued with the trucks and, after 1918, automobiles were assembled by special order only.

The **Cleveland Motor Car Company** was founded in 1904 by E. J. Pennington. Before founding the car company, from 1900 to 1902, Pennington sold the *Tractobile*, which was a steam device that could be attached to a carriage to turn it into a steam-driven vehicle.

Once the company started manufacturing their own autos, the autos were in the touring car style. The 1906 *Cleveland Model F* had a four-cylinder, 18-horsepower engine, selling for $2,800. Two body options were available: Tulip and Victoria. They also built a 20-horsepower *Model D* at the same time. A larger version, selling for $4,000, was not competitive compared to other cars in the price range. The **Cleveland Motor Car Company** went out of business in 1909.

The 1907 Cleveland

Two other cars were called the *Cleveland*; one was advertised as "the Aristocrat of Cyclecars" in 1914. It had a water-cooled, four-cylinder engine and seated two people side by side. The other *Cleveland* auto was an electric car made from 1899 to 1901, assembled by the **Cleveland Machine Screw Company**. The vehicle was a light electric two-seater with rubber tires, also known as the *Sperry*.

The *Ewing* of Geneva, Ohio, was a taxicab manufactured in 1908, designed by Louis Mooers and named after the largest investor of the **Cleveland Auto Cab Company**, Levi Ewing. As his investment grew, Ewing gained control of the company in February 1909 and renamed it the **Ewing Automobile Company**.

The car itself was a four-cylinder, 24-horsepower vehicle available as a taxi or as a town car, selling for $3,000. About one hundred were made when the company was purchased by Billy Durant for **General Motors** in 1911.

A big factor contributing to Ewing wanting to sell the company was that he said he would have to move the company to a larger urban area otherwise. He was unable to maintain a stable workforce in small-town Geneva, Ohio, since the town had no nightlife and "nothing to do."

Walter Baker of Cleveland founded the **Baker Motor Vehicle Company** in 1898. The first cars by the company could go seventeen miles per hour and travel fifty miles on a single charge. Future vehicles included a 1910 limousine. One of the first customers to buy one was Thomas Edison. A *Baker Electric* was chosen by President Taft to be part of the White House fleet.

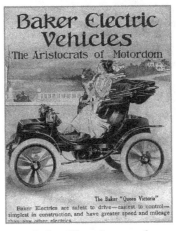

Baker Electric car ad

In 1915, the **Baker Motor Vehicle Company** purchased the **R.M. Owen Company**, who made another electric car, the *Owen Magnetic,* and then merged with the **Rauch & Lang Company**. The last car with the *Baker* name was in 1916, but **Rauch & Lang** continued manufacturing electric autos until the late 1920s.

1915 Rauch & Lang auto

The Peerless Manufacturing Company made clothes wringers before switching to autos in 1900. The Cleveland-based company had also done some bicycle manufacturing before turning to autos. The first venture into autos was achieved by putting a one-cylinder motor in a buggy and calling it a *Motorette*. Chief engineer Louis P. Mooers joined the firm in 1901, and by 1904, they had a four-cylinder model ready to exhibit at the New York Auto Show.

Some of the features of the early **Peerless** autos included a tilting steering wheel, allowing easy entrance and exit from the vehicle, and one of the first side entrances for a vehicle. In 1902, ninety cars were built, and the company name was changed from the Peerless Manufacturing Company to the **Peerless Motor Car Company.**

In 1904, the company released the *Peerless Green Dragon* with 60 horsepower, a car which driver Barney Oldfield used to set many speed records. With the slogan "All That the Name Implies," they built over one thousand cars a year by 1906. The brand had a sterling reputation for mechanics and styling and was considered one of the three "P"s of luxury cars along with *Pierce-Arrow* and *Packard.*

The six-cylinder vehicle the company brought out in 1907 was popular, as was the V-8 engine introduced in 1916. The company went through many management upheavals but stayed the course, producing upward of over four thousand per year of the luxury marque.

In 1925, the **Peerless Motor Car Company** management decided to produce a smaller car that was inexpensive, geared to the general public. With the slogan "Now There's a *Peerless* For Everybody," they released a six-cylinder car selling for $1,500. The new V-8 engine in the regular-sized *Peerless* was popular, even though the $4,000 price tag was a bit steep.

1917 Peerless Model 56 seven-passenger touring car

A new style for the *Peerless* in 1930 was done in vain as the nation went into the Great Depression. As the nation got poorer, it was the worst time possible for a luxury car company to bring out expensive new models, and *Peerless* went under in 1931. The company had been ready to release a car with a V-12 engine when the Depression hit. Instead, the last year for *Peerless* was 1932, when the models built in 1931 were sold.

When they closed their doors for good in 1932, it was the end of Cleveland and Ohio as a major auto maker, although many auto-parts companies are still headquartered there.

Peerless Auto Factory and Headquarters in Cleveland, built in 1908

Arthur L. Garford made a fortune in 1892 on his "Garford seat," the first padded bicycle seat. In 1909, he established the **Garford Motor Truck Company** in Elyria, Ohio. From 1904 to 1911, they had a partnership with the **Studebaker Motor Company** to make automobiles, the *Studebaker-Garford*. In 1912, they received a very profitable order from the U.S. Post Office for delivery trucks.

Studebaker-Garford automobile

In 1925, the company moved to Lima, Ohio, and changed its name to the **Superior Body Company**. They had a lucrative business selling hearses, ambulances, and school buses in partnership with **Studebaker**. In 1940, they changed their name to the **Superior Coach Company**. They were purchased by the **Sheller-Globe Corporation** in 1969. In the late 1980s, the company had big losses in their school bus and hearse lines, and consequently, portions were liquidated by the parent corporation. **Superior Coach Company** still makes bodies for various vehicles.

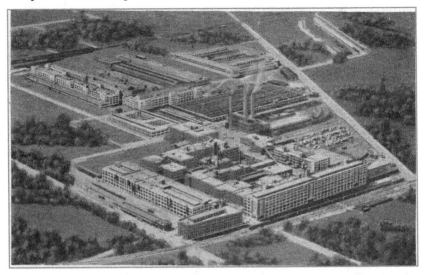

Factory of the Overland Motor Company

Claude Cox was working for the **Standard Wheel Company** of Terre Haute, Indiana, when he developed the prototype that became the first *Overland* auto in 1903. It was one of the first hybrid vehicles, having a two-cylinder, water-cooled, gasoline-powered engine and a switch to change power to two batteries. In 1905, Cox moved the company to Indianapolis and manufactured the *Overland* until 1908.

In 1908, when John Willys came to the company to find out what had happened to the cars he had ordered for his car dealership, he found out the company was going into receivership the following Monday.

Willys purchased the company himself and continued producing the *Overland* auto until 1926. In 1912, the company name was changed to the **Willys-Overland Motor Company**. The company was number two in auto sales from 1912 to 1918 (**Ford** was number one).

In 1916, Willys purchased what was then Canada's most successful car company, the **Russell Motor Car Company**. Another acquisition was in the mid-1920s when he purchased the **F.B. Stearns Company** and continued making their popular *Stearns-Knight* auto. A *Willys* outfitted with the Knight engine was called the *Willys-Knight*.

The *Willys Whippet* replaced the *Overland* in 1926. At the dawn of the Great Depression, the *Stearns-Knight* was discontinued in 1929. The *Whippet* stopped production in 1931, and 1933 was the last year for the *Willys-Knight*. Autos called the *Willys Six* and the *Willys Eight* were manufactured instead, and in 1936, the *Willys 77* debuted.

The contract to build **Willys's** most enduring brand, the *Jeep*, was negotiated by Joe Frazer (later to be part owner of the **Kaiser-Fraser Corporation**) in 1938. The company started manufacturing the *Jeeps* for the government in 1941. After the war, they continued producing *Jeeps* but discontinued their other brands.

In 1952, they re-entered the private market with the *Willys Aero*, a passenger car that was a six-cylinder, two-door sedan.

In 1953, the company was purchased by the **Kaiser Motor Company**. The new company continued to manufacture *Aeros* for a few more years and then concentrated on *Jeeps*, even changing the name of the company to the **Kaiser-Jeep Corporation** in 1963.

The **Kaiser-Jeep Corporation** was purchased by the **American Motors Corporation** in 1970. **AMC** continued making *Jeeps* until they were purchased by the **Chrysler Corporation. Chrysler** is still producing the *Jeep* today.

———————————

The **Elmore Manufacturing Company** got its name from its town: Elmore, Ohio. The company started making bicycles in 1893 and in 1904 manufactured the *Elmore* auto, a two-cylinder convertible with attachable top (tonneau) as an option. Production was moved to the nearby village of Clyde at 504 Amanda Street, where auto production began. The company was owned and operated by Harmon Becker and his sons, James and Burton. The company successfully operated for eleven years, then Billy Durant purchased the company to become part of **General Motors.** When Durant was forced out in 1910, support for the auto dwindled, and the car became defunct in 1912.

A 1919 Holmes Series A

Another bicycle builder that switched to making autos was the **Kirk Manufacturing Company** of Toledo, Ohio. Although they made the announcement about the switch in 1899, it wasn't until 1902 that their car hit the market. It was called the *Yale* and had a water-cooled, two-cylinder runabout with a detachable tonneau (top), selling for $1,500. It had a chain drive and planetary transmission. All parts except the differential and the oil device were made "in-house." Their slogan was "The Car With the Doubt and the Jar Left Out."

Business was good for the 1902 year, with a Chicago distributor selling fifty units. In October 1903, the **Kirk Manufacturing Company** consolidated with the **Toledo Manufacturing Company** and another bicycle company, the Shell Cycle Fittings Company. With the added capitalization, they built a new factory in Toledo. Their new name was the **Consolidated Manufacturing Company**.

Below: Yale auto advertisement, Yale logo

In 1904, the *Yale* came out with two touring models of the two-cylinder cars and in 1905 brought out a larger, four-cylinder car, selling for $2,500.

Maybe it was too much, too soon, for the new, consolidated company announced it would be selling their automobile division and concentrating on bicycles and motorcycle and auto parts. By May 1906, they filed for bankruptcy.

The *Columbus Electric* was built in Columbus, Ohio, and was an electric auto built by the **Columbus Buggy Company** in 1903. It could achieve speeds of twenty-one miles per hour and seventy-five miles per charge. They had started making buggies in 1875. In 1907, they started making gasoline autos under the name *Firestone-Columbus*, along with electric cars.

Eddie Rickenbacker driving presidential candidate William Jennings Bryan in a 1909 Firestone-Columbus touring car

The *Stoddard-Dayton* auto was another brother act—in this case, John and Charles Stoddard. The family business was at first making hay rakes in Dayton, Ohio, for horses in 1875. By the time the **Stoddard Manufacturing Company** was incorporated in 1884, the family was also making bailing forks and a variety of other farm implements.

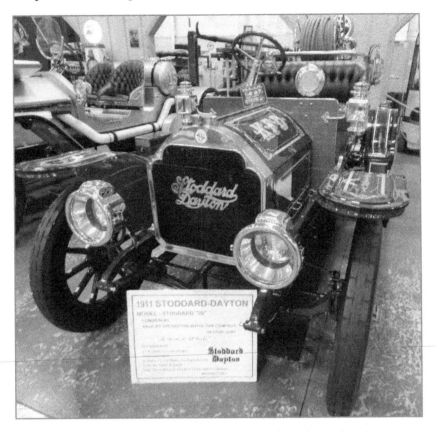

1911 Stoddard-Dayton

The younger brother, Charles, had become interested in autos during a trip to Europe in the 1890s. He began experimenting, and in 1904, the first *Stoddard-Dayton* came out with a four-cylinder, 26-horsepower engine and attractive body. By December, the name of the company was changed to the **Dayton Motor Car Company**, and cars replaced farm implements in the Stoddard family factory.

In 1905, the motto "As Good As It Looks" aptly described the *Stoddard-Dayton*, and it was considered by car critics to be "large, luxurious, and sturdily built." Production started at about five hundred in 1905 and 1906 to over two thousand autos manufactured in 1907. In 1908, a car with six cylinders was introduced. By this time, the company was offering models ranging from coupes and runabouts to roadster, touring, and limo models.

The *Stoddard-Dayton* was doing well, but it was expensive. In 1909, it was announced that a less expensive car would be made by a subsidiary, the **Courier Car Company**. The next year, 1910, the company made the fatal mistake of joining Benjamin Briscoe's conglomerate, the **United States Motor Company**.

In 1911, the *Stoddard-Knight* model came out, as the company was the second (after **Stearns**) to sign a contract with engine maker Charles Knight to use his engine. In 1912, a *Stoddard-Knight* six-cylinder, 70-horsepower engine was available on the *Stoddard-Knight*. In 1912, the **Courier Motor Company**, also a division of the conglomerate, called their cars *Courier Clemons*. In 1913, however, the company was out of business, along with the rest of the companies of the **United States Motor Company** except for the **Maxwell Motor Company** (which evolved into **Chrysler**). The other discontinued companies included the **Columbia Automobile Company**, the **Brush Motor Car Company**, and **Alden Sampson Trucks**. Also on the chopping block was the **Dayton Motor Company** subsidiary, the **Courier Motor Company**.

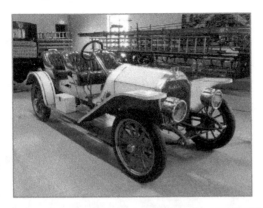

1910 Stoddard-Dayton and the 1909 Courier Model 10-A-1

Although the **Ben-Hur Motor Company** was incorporated in Delaware in 1916, most of the manufacturing took place in Willoughby, Ohio. Founded by L. L. Allen, the auto had a six-cylinder engine, a "forced-feed" oil system, and a 126-inch wheelbase and came in touring and roadster models. It had a three-speed transmission and magneto ignition. The small factory was limited to making about twenty cars a week and announced that due to the difficulty in securing chassis, production would actually be less than that number. The lineup for the cars included a five- to seven-passenger touring car for $1875, a seven-passenger sedan for $2750, and a four-passenger roadster for $1875.

Even though 1918 models were announced, by May of that year, the company was in receivership and disappeared entirely shortly thereafter.

A Ben-Hur Motor Company stock certificate and radiator badges

Chapter 36:
The Tulsa and the Geronimo of Oklahoma

WHERE WAS THE AUTO KNOWN as the *Tulsa* built? Go to the head of the class if you guessed Tulsa. The *Tulsa* was manufactured in Tulsa, Oklahoma, by the **Tulsa Automobile Company**. With limited production, they made cars with four and six cylinders. They were in business from 1917 to 1923. The *Tulsa Four* was a four-cylinder, 35-horsepower auto, and a six-cylinder version was added in 1921. The factory produced ten cars a day. These were sold in the Tulsa and Oklahoma City areas.

The 1920 Tulsa, an "oil field runabout."

The *Geronimo* was an open-modeled car that started manufacturing a four-cylinder auto named after the Indian chief in 1917 at 409 S. Grand Street in Enid, Oklahoma. The founder was William C. Allen. Their two models were the *4A-40*, a roadster capable of 45 horsepower, and an open touring car seating five and with 55 horsepower called the *6A-45*.

At the end of 1917, the company felt successful enough to build a thirty-thousand-square-foot factory on the outskirts of Enid. They sold their autos mostly in the Midwest states of Kansas, Nebraska, west Texas, and locally in Oklahoma. They also sold tractors. In 1918, the *Geronimo* switched to a six-cylinder engine. At various times, they employed between forty and 125 workers.

In 1920, the company's plant burned and caused $250,000 in damage, of which only $65,000 was covered by insurance. This caused the company to fail. The company produced about one thousand autos in its lifetime.

1918 Geronimo

In El Reno and Oklahoma City in Oklahoma, the *Pioneer* auto was produced by the **W.R.C. Motor Works** from 1907 to 1912. The vehicle was a high-wheeler model with a two-cylinder engine capable of 20 horsepower on a ninety-one-inch wheelbase with a planetary transmission and shaft drive.

By 1909, the company had sold only twenty-two cars. They incorporated at $10,000 and renamed the venture the **Pioneer Auto Company,** but when the *Pioneer* made a good showing in the Oklahoma City–Kansas City endurance run, only losing to a *Mitchell* in the final hours of the race, the company gained more orders than they could fill. This included fifty orders for new autos from the Wichita Auto Show that year.

They became so popular that they received offers to move the factory, including an offer by Wichita, Kansas, and then Oklahoma City, an offer they accepted. Unfortunately, the going was not so good in Oklahoma City, despite the four-cylinder car they came out with in 1910. In 1911, they sold out to the **Tulsa Auto and Manufacturing Company**.

The Pioneer automobile was produced from 1907 to 1912

Chapter 37:

Don't Give the Diehl Motor Company of Oregon Any of Your Money!

F NOT FOR G. H. Diehl, the Portland, Oregon, designer for the 1935 *Diehl*, Oregon's chapter might have been a blank page.

The *Diehl* was a teardrop-shaped car with a streamlined and fenderless body and the engine in the rear. It had two wheels in the rear and one in the front—a quality that Diehl claimed made steering and parking easier. The car had a single headlight, and only about five were ever made.

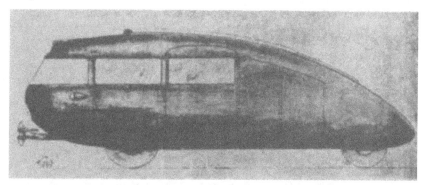

1935 Diehl of Portland, Oregon

Chapter 38:

The Transient Car Companies of Pennsylvania

PENNSYLVANIA HAD SEVERAL CAR COMPANIES that started in one part of the state and moved to another.

The **Pittsburgh Motor Vehicle Company** of Pittsburgh, Pennsylvania, started in 1897. They built what were at first experimental vehicles designed by Lewis S. Clarke. The first one was a three-wheeled, gasoline-powered "tricycle," which is in the Smithsonian Institute in Washington, D.C. Lewis (or Louis) Semple Clarke is also credited with inventing a spark plug, a drive-shaft system, and an oil-circulation system. He also started putting the steering wheel on the left side, leading to America driving on the right-hand side of the road.

The 1910 Autocar Roadster and the Autocar Radiator badge

In 1899, the company moved to Ardmore, Pennsylvania, and changed its name to the **Autocar Company**. In 1902, they began production of a vehicle that is believed to be the first two-cylinder vehicle with shaft drive.

*Autocar*s were popular and considered well made. In 1901, they were fifth in the nation in motor car sales. From 1901 to 1911, they manufactured about 5,800 vehicles. They advertised that it was the car that "cannot blow up or burn up," no doubt addressing the fears that some people had about gasoline engines at the time. Steam- and electric-car companies were quick to stir up controversy over gasoline engines.

The company started to make trucks in 1907, and by 1911, the company dropped the cars to make trucks exclusively, although continuing to call themselves the **Autocar Company**. In 1953, the company was purchased by the **White Motor Car Company**. **White** was taken over by **Volvo Trucks** in 1981, and *Autocar* was maintained as a separate division, until it was sold off to the GVW Group, LLC in 2001, when it was revived as a separate company to make heavy trucks.

1904 Autocar

In 1894, James T. Reber founded the **Acme Machine Company** in Reading, Pennsylvania. They manufactured bicycles and cigar-making machinery. His son, James C. Reber, and the **Reber Manufacturing Company** built the 1902 *Reber,* a two-cylinder, 20-horsepower, five-passenger, rear-entrance tonneau model car.

In 1903, Reber changed the name of the company to the **Acme Motor Car Company**, and in 1911, Herbert M. Sternbergh, Robert E. Graham, and Fred Van Tine formed an auto company in Reading, Pennsylvania, using their initials, **SGV**. Engineer Van Tine designed a four-cylinder, shaft-driven car inspired by the Italian sports car *Lancia*. In fact, the car was sold by the same New York dealer that sold the *Lancia*.

The *C.B.G.* model included a hot-water-jacketed intake manifold and a system to force-feed lubrication to the engine bearings and pistons. The car had classy features, including Circassian Walnut dashboards and custom coachwork. At their peak, the company built thirty-five to forty of these autos per month.

The models were not only popular with the American wealthy, including the Astors and the Vanderbilts, but **SGV**s made it across the pond to the garages of the King of England, the Kaiser of Germany, and the Czar of Russia. With all the extras and customizations, many **SGV**s would cost $12,000 or more.

Life was rosy until a Vulcan four-speed electric gearshift, which was mounted in the wheel spokes and activated by a button on the steering wheel, malfunctioned in a whole shipment of autos. This incident created bad publicity for the company, which in 1915 was already on shaky financial ground. By the end of the year, the company was sold at auction, including thirty-two autos in various stages of completion.

1912 SGV roadster

In 1911, the **Nance Motor Car Company** came out with its six-cylinder, 35-horsepower vehicle. It had a 122-inch wheelbase, was available in a roadster or touring model, and sold for $1,900. The following year, the car name and company were changed to *Touraine* to reflect a new financier. The man in charge continued to be Harold B. Larzelere.

The **Touraine Motor Company** then produced a six-cylinder auto in Philadelphia in 1912, which was basically the same as the *Nance* in 1911. In 1913, phaeton models were added to the line; in 1914, limousine, raceabout, runabout, roadster, and phaeton models were available.

In 1915, the company decided to make the 1916 models the last passenger autos produced and to concentrate on commercial vehicles.

The **Touraine Motor Company** of Philadelphia started making commercial vehicles under the brand name of *Vim* in 1913. (In 1914, they announced a cyclecar with the *Vim* name, although whether it reached the production stage is not known.) The company name was changed to the **Vim Motor Company** in November 1915. Trucks continued to be produced under the *Vim* name until 1923.

The *Twyford* was built by the **Twyford Motor Car Company** in Pittsburgh and Brookville, Pennsylvania, from 1899 to 1907. The founder, Robert E. Twyford, was a big proponent of four-wheel drive, and his cars were considered the first. The rear-mounted, two-cylinder-engine vehicle had a crude, early version of power steering using bevel gears.

1904 Twyford Stanhope

The first car was called the *Twyford Stanhope* and sold for $1,200, but slow and low production caused them to accept an offer from Brookville, Pennsylvania, to build a factory on a ten-acre site there. The one they opened in 1904 was a two-story, brick building. However, the company was gone by 1907. Robert Twyford continued to profit through his four-wheel drive patents.

*The Twyford factory of Brookville, Pennsylvania,
with various Twyford autos parading through town*

The original *Austin* auto was designed by Sir Herbert Austin in England in 1922. It was a much smaller car than usual and took off in Europe. By 1924, *Austin* establishments had opened in Japan, France, and Germany.

By 1930, an American branch was opened. The *American Austin* was produced by the **American Austin Motor Car Company** from Butler, Pennsylvania, from 1930 to 1934. It was a subcompact and had fixed disc wheels with detachable rims and "Chevrolet-like styling." It sold 8,558 cars in its first year, which was below expectations.

With some lean years, including going into receivership in 1932 and 1934, they renamed the car the *Bantam* in 1935 and the company the **American Bantam Car Company**. The car was improved with pressure lubrication, a mechanical pump feed, and a new grille with horizontal bars.

American Bantam logo and 1929 Austin Seven

Because of the small size of the auto, jokesters would often make an *Austin* owner sorry he bought one. Students would often take the cars and leave them in funny places such as the principal's office. However, during the war years from 1941 to 1945, the *Bantam*s and *American Austin*s were extremely popular due to their gas economy and minimal maintenance costs. Unfortunately, they were no longer producing autos.

In 1940, they offered larger, 800cc engines and three-bearing crankshafts. The models offered included a four-seater convertible, a touring car, a roadster, and a station wagon. By 1941, they started making *Jeep*s for the military instead of passenger cars. After producing 1,500 *Bantam Jeep*s, the U.S. Army decided to go with the **Willys** design instead, and they shipped all the *Bantam Jeep*s to Russia.

1940 American Bantam

In Bryn Mawr, Pennsylvania, the **Pennsylvania Auto Motor Company** was organized in 1907. They built a car called the *Pennsylvania* in several styles. This included four- and six-cylinder cars of 29 to 75 horsepower, ranging in price from $2,100 to $4,700. In 1911, creditors, including **J.M. Quinby & Company**, the firm that made the cars' bodies, sued the company into receivership.

Pennsylvania Auto Company advertisement

CHAPTER 39:

FROM GROCERIES TO CAR COMPANIES IN RHODE ISLAND

N 1896, THE *BALDWIN STEAM* was built in Providence, Rhode Island. The first vehicle they built was a steam delivery van for a local drugstore. Next, in 1899, was a steam-powered, four-passenger vehicle with a two-cylinder engine and a top speed of twenty-five miles per hour. Calling themselves the **Baldwin Automobile Company** and then reorganized as the **Baldwin Motor Car Company**, they moved to Connellsville, Pennsylvania, in 1900 and manufactured a small number of vehicles.

In 1902, the company developed severe financial difficulties and was sold at auction to a consortium that included Queen Liliuokalani of Hawaii. The group, named the **American Locomotor Manufacturing Company**, had plans to move operations to Morgantown, West Virginia, but these plans never came to fruition.

In 1901, J. P. Thompson, who was originally in the grocery business, started the **Thompson Motor Company** in Providence, Rhode Island, to manufacture trucks, buses, and delivery wagons. The *Thompson* was steam powered, and enough were produced to keep the company afloat until 1907.

From 1909 to 1913, the *Alco* was an auto built in Providence, Rhode Island, by the **American Locomotive Company**. The car could be bought customized—and a customized *Alco* won the Vanderbilt Cup in 1909 (the car's first year) and in 1910. The high price—$6,000 to $7,500—was a point of pride with the company, who bragged that they were the most expensive car. When the company figured out that they lost $450 on each custom job, they quit making the auto.

———

In 1903, Everett S. Cameron and his brother Forrest assembled the air-cooled cars known as the *Cameron*. The first few were steam vehicles, while later models moved on to gasoline engines.

The first non-steam car they built was in 1900. This first gasoline car had one cylinder. The autos revolutionized torque tube drive and wheel steering. They stayed in business until 1920 as the **Cameron Manufacturing Company** with plants in New London, Connecticut, and Attica, Ohio. They had a truck factory in Alma, Michigan.

The Cameron

Cameron auto display

Chapter 40:

The Anderson of South Carolina: Built in Dixie!

THE **Anderson Car Company** of Rock Hill, South Carolina, was considered the most successful car company in the American South. (One of its mottos was "A Little Higher In Price, But It Was Made In Dixie!") The cars were sold in other parts of the country as well. Operating from 1916 to 1926, they successfully manufactured a popular, well-made car that was well marketed. They featured several different models, body designs, and unique color schemes.

In 1889, the *Anderson* got its start in the back of a furniture store in Rock Hill, South Carolina, as the **Holler and Anderson Buggy Shop**. First centering on repairs, the company became the **Rock Hill Buggy Company** and began manufacturing their own horse-drawn carriages and buggies.

The 1923 Anderson touring car

John Gary Anderson was one of the founders and began to see the handwriting on the wall with the advent of motor vehicles. He tried producing a car in 1910 but failed to start a company. For the next two years, the buggy company was successful and adding multiple shifts. By 1912, however, Anderson was tweaking his model, and by 1916 began changing his business from carriages to cars. He raised $1.5 million through public stock offerings, changed the company name to the **Anderson Motor Car Company**, and began to produce the self-named *Anderson*. Like most carmakers of the times, he accomplished this by buying most of the parts from suppliers, which he assembled into his own unique design.

The first *Anderson Six* auto had 40 horsepower and a six-cylinder engine. The first one was delivered to the president of Rock Hill's bank. The car had a lot of "extras" that attracted customers. Some of these were detachable side curtains for the windows (with an envelope to put them in), a ventilating windshield, a cigar lighter, a power-driven tire pump, a complete set of tools, and even the running boards included luggage straps. No wonder one of their mottos was "It Sells Itself!" The standard paint job color was Brewster Green, with black fenders and hood. Other color combinations were available for an additional price.

Successful auto companies were desired, and Anderson refused an offer of $1 million to move the company to Charlotte, South Carolina. John Anderson remained loyal to Rock Hill and refused this, and all other, offers.

In 1920, the company manufactured 1,180 autos. By 1923, that number had risen to 1,875. Also in 1923, the company manufactured what was advertised as "the most inexpensive, all-aluminum car." Unfortunately, in 1924, engine problems were common on the company's *Model 41*, and in the same year, a factory fire was responsible for $40,000 worth of damage. Their eventual end came when their $2,000 price couldn't compete with the **Ford** *Model T*'s $200 price.

In the ten-year life of the auto company, over ten thousand *Anderson*s were produced.

"It's As Roomy For Five As It's Chummy For Two" was one of the *Anderson* mottos.

Chapter 41:

The Fawick Flyer and Reese Aero-Car: Flying High in South Dakota

In 1883, E. Scott Callihan of Woonsocket, South Dakota, built a two-cylinder, three-wheeled steam car. From 1883 to 1905 he assembled, by hand, several different steam cars. He built the whole car, even the parts most companies buy preassembled. The top speed for his cars was fifteen miles per hour. Callihan was one of the first in the Dakotas to have electricity. He constructed his own power plant in his backyard.

———

The **Silent Sioux Auto Manufacturing Company** began in Sioux Falls, South Dakota, in 1908. It was formed to produce the car made by Thomas L. Fawick when he was only eighteen years old, the *Silent Sioux*. The cars were assembled on West 13th Street in Sioux Falls.

In 1909, the company name was changed to the **Fawick Motor Car Company** and the name of the car to the *Fawick Flyer*. This was a large touring car and had a 40-horsepower, four-cylinder Waukesha engine. It also had an aluminum body and could go over sixty miles per hour, even though the speed limit in Sioux Falls was only seven miles per hour. The *Fawick Flyer*, it was claimed, was the first car with four doors, and it was the car that Teddy Roosevelt rode in when he visited Sioux Falls. The cars sold for $3,000, eliciting a $600 profit. Production was limited to fewer than ten.

Teddy Roosevelt (on the right) in a Fawick Flyer in Sioux Falls in 1910. This car was considered to be the first four-door auto.

The company went defunct in 1912, but Fawick went on to find gainful employment in the auto industry. For a while, he operated a business that assembled the airflex clutch he invented. After World War II, he purchased the **Federal Motor Truck Company**. At the time of his 1978 death, he operated the **Fawick Corporation**, which sold his patented industrial brakes and clutches.

Fawick had over 250 patents, and besides automotive, he also had patents for industrial machines and invented the first rubber-gripped golf clubs. He was also a musician and built violins and sound systems. He wrote music and collected art. His art was worth $1.2 million upon his death, and most of the collection was purchased by the Cleveland Museum of Art.

Thomas L. Fawick

Fawick in 1955 with a restored Fawick Flyer

CHAPTER 42:

THE MARATHON AND DRAKE
OF TENNESSEE

TENNESSEE IS CONSIDERED THE MAIN state in the Southern Automotive Corridor, with three main assembly plants and the **Nissan** headquarters within its boundaries. They are now a big "car area," so it would be ironic if no early car companies existed within the Tennessee boundaries.

Marathon advertisement

1907 Marathon auto in front of Marathon factory

No irony here, since the **Marathon Motor Company** was based in Nashville, Tennessee, and manufactured cars from 1911 to 1914. They grew out of an earlier company called Southern Engine and Boiler Works in 1889. Their car, the *Marathon,* came in roadster and touring models with four-cylinder, 35-horsepower engines, selling for about $1,500. The car was originally named the *Southern* and produced about six hundred cars under that name in 1906 until another auto-manufacturing firm was found to have prior rights to the name.

The **Marathon** factory was originally built as the Phoenix Cotton Mill. It was expanded as needed—**Marathon** soon had dealers across the country.

The Marathon Motor Works factory, built in 1881,
now the site of antique businesses

The car was successful, and by 1911, five models were offered. By 1912, they were producing two hundred cars per month. In 1913, the company produced twelve different models, but management trouble behind the scenes caused the company to fail in 1914.

In 1921, the **Drake Motor & Tire Manufacturing Company** of Knoxville, Tennessee, was capitalized at $3 million. The founder, W. F. Drake, proposed that the company manufacture and sell autos, auto bodies, trucks, tractors, sheet-metal parts, tires, and rubber by-products. While they never manufactured much, what they ended up making were several four- and six-cylinder touring cars and a five-passenger sport model named the *Drake*.

A custom-built, seven-passenger limo model with a hood of polished aluminum was built for silent screen star Mildred Reardon, who provided good press for the auto by posing with it for publicity shots, but it was too late to save the brand, as the *Drake* succumbed to the postwar Depression in 1922.

Silent film star Mildred Reardon by her custom-made Drake 6. She appeared in Hal Roach and Cecil B. DeMille pictures. She died young at the age of thirty-seven after a life marred by drug addiction.

Chapter 43:

The Tex, Texan, and Texmobile:
The Texas Car Companies

ANY OF THE TEXAS CAR companies had some form of the word "Tex" within their titles: the *Texan*, the *Texmobile*, the *Tex*, and even an out-of-state steam car from Orange, Massachusetts, called the *Texas*. Following the theme of Texas heritage was also the *Dixie*. Probably the earliest attempts at mass producing an auto in the Lone Star State was in 1905 and 1906, when the **Hawkins Automobile & Gas Engine Company** built the *Hawkins* in Houston, Texas.

One of the first, longest-lived, and most successful of the Texas auto companies was the **Wichita Falls Motor Truck Company.** The company started in 1911 in Wichita Falls, Texas, and was in business until 1932. They built heavy-duty *Wichita* trucks.

Wichita Trucks were very well regarded for their durability and strength. They made trucks for the logging industry, tankers for the oil industry, off-road vehicles, and military and fire vehicles. During World War I, they manufactured 3,500 vehicles for France and over four hundred for Russia.

They produced the *Wichita Combination Car*, a light-duty vehicle, from 1920 to 1921, but it was unsuccessful. More successful were its heavy-duty dump trucks as state and federal road building projects used the trucks all through the 1920s. One of their last trucks was one of the first cement mixers, manufactured in 1930.

Like so many other companies, the **Wichita Falls Motor Truck**

Company went out of business during the Great Depression in 1932. They made over ten thousand vehicles over the company's lifetime and sold them around the world. Countries they exported to include Belgium, Burma, Chile, China, Costa Rica, Cuba, Ecuador, England, France, Gold Coast (Australia), Greece, Holland, India, Iran, Italy, Jamaica, Japan, Mexico, Morocco, New Zealand, Panama, Russia, Spain, and Thailand. The factory of the **Wichita Falls Motor Truck Company** on Arthur Street is now the Wichita Energy Company, and they have a collection of *Wichita* trucks.

Wichita Falls Motor Truck Company logo

Early Wichita commercial vehicle

Several companies built a prototype, or one or two models, and not much more was heard from them. From 1915 to 1922, the **Blumberg Motor Car Company** built the *Blumberg* in Orange and San Antonio, Texas. The **Lone Star Motor Car Company** operated from 1919 to 1922 at 515 Roosevelt Road in San Antonio, Texas. They were also known as the **Lone Star Truck & Tractor Association**. In 1921 in San Antonio, the **Robertson Company** built a car named the *Robertson*. In 1909, the *McGill* was built by the **McGill Motor Car Company** of Fort Worth, Texas. It was a short-lived, six-cylinder car with an engine capacity of 3450cc and cost $2,385. From 1914 to 1916 in Waco, Texas, the *Hall Cyclecar* was built by the **Hall Cyclecar Manufacturing Company**. A Houston company named the **Southern Motors Company** built the *Southern Six* from 1920 to 1924 and a truck named the *Ranger* from 1920 to 1923. In 1918 in Fort Worth, Texas, the **Bridges Motor Car & Rubber Company** built the *Bridges* auto.

CHAPTER 44:

THE "HERE TODAY, GONE TOMORROW" ROYALMOBILE OF UTAH

IMAGINE YOU LIVE IN UTAH, and you have just purchased this book in the Salt Lake City International Airport to read on your flight. You want to read about the auto companies of your home state and have come to this chapter, but unfortunately, a car with the unlikely name of the *Royalmobile* is the sole car listed for Utah. Even more unfortunately, the *Royalmobile* (or *Royal Mobile*) was mostly a scam that only started in Utah and then moved to Ashland, Ohio, making your state one of only two (see also Nevada) that not only had only one car company, but that one was phony!

H. K. Clover breezed into Salt Lake City, Utah, in 1901 and claimed to have over two hundred patents to his credit. He had an idea for a steam car with a "compound engine and chain" and began collecting shares in a company that he formed to make his auto, the *Royalmobile*. Due to the "auto craze" of the 1890s–1900s, when everyone thought that starting a successful car company was the way to riches, Clover sold a lot of shares.

In addition, Clover managed to work out a deal where he got a four-acre lot and a $15,000 bonus from Hastings, Ohio, to move his factory there. In return, Clover promised that he would build a factory that would employ at least one hundred men for at least four years. This, of course, never happened, and now, whoever got this book at the airport to read about Utah's car industry will have to settle for the story of the not-so-*Royalmobile*.

Chapter 45:

The Vermont Car Company that Built the Wasp

VERMONT, THE STATE THAT FIRST instituted a "red-flag rule," requiring a pedestrian to walk one hundred feet in front of a vehicle in warning, hosted at least one auto company after they relaxed their rules somewhat.

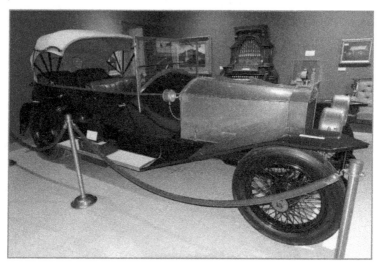

The 1924 Wasp of Vermont

The most notable one of these was the *Wasp*, built in Bennington, Vermont, by the **Martin-Wasp Corporation** from 1919 to 1925. Kar Marin designed the *Wasp*. A total of eighteen *Wasp* models were manufactured over the life of the vehicle. The earlier models had four cylinders and Wisconsin T-head engines. Later cars used a Continental engine. Every *Wasp* had a St. Christopher medal on its dashboard!

Chapter 46:

Traveling in the Piedmont in Virginia

THE FIRST VIRGINIA-MADE CAR WAS the *Dawson*, a basic car built in Basic City (now Waynesboro) in 1901.

The **Piedmont Motor Car Company, Inc.,** started in 1917 in Lynchburg, Virginia. They manufactured four-cylinder cars with Lycoming engines and six-cylinder cars with Continental engines in various open and closed body styles.

Piedmont also assembled cars for other companies, including **Bush** (Chicago), **Alsace** (basically a *Piedmont* with a *Rolls-Royce* side radiator, built for **Automotive Products Company** in 1920 and 1921), and **Lone Star** (of San Antonio, Texas, 1920–1922). **Piedmont** would assemble the car and then put the appropriate radiator symbols, badges, and other identifiable markings on. The *Piedmont*s themselves were usually "Piedmont Green." With the outside work, at their height, the **Piedmont** factory was producing about five hundred cars per year.

They made their outside car-assembly work a priority, so sometimes, when a customer wanted a *Piedmont*, one wasn't ready, which was not good for business. **Piedmont** also overstocked on parts, causing them to go into receivership in 1922 with over $225,000 in remaining parts still in stock.

James Kline was persuaded by Virginia businessmen to move his Pennsylvania based company **BCK Motor Company** to Richmond, Virginia. From 1912 to 1923, he produced approximately 3,700 *Kline Kars* in a new factory built in Richmond. The cars consisted of many different models, with prices from $2,000 to $3,500, expensive for the day.

The Kline Kar of Virginia

Chapter 47:

The Washington Hydromotor Could Go on Land or Sea

I N 1914, WILLIAM MAZZEI LED the project of the **Automobile Boat Manufacturing Company** of Seattle, Washington. They built an automobile that would double as a boat, probably the first working amphibian car. It would go sixty miles per hour on land and twenty-five miles per hour in the water. When the vehicle entered the water, the drive wheels locked, while the engine powered the propeller. The steering wheel operated the rudder and the front wheels.

In the 1950s, in the early days of television, a children's show was inexpensively produced about *Supercar,* with the characters portrayed by marionettes, but the car itself, when it had its own scenes, was elaborately produced. The car could go on land and sea! Was the 1914 amphibious *Hydromotor* the inspiration for the 1950s *Supercar*? Well, full disclosure time ... the *Supercar* could also *fly*!

Although the vehicle was well received, the high price did not produce enough sales for the **Automobile Boat Manufacturing Company** of Seattle to continue beyond 1917.

The Amphibian Hydromotor car and
the 1917 Hydromotor Amphibian car

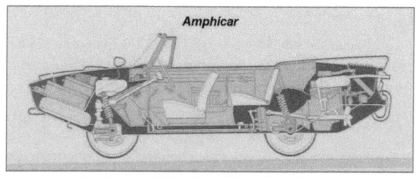

Design model for Hans Trippel's Hydrocar

CHAPTER 48:
THE NORWALK OF
WEST VIRGINIA

THE NORWALK MOTOR CAR COMPANY, manufacturers of the auto known as the *Norwalk,* didn't start in West Virginia—they were founded in Norwalk, Ohio, in 1910. Their first car was underpowered with a 251-cubic-inch engine but only had four cylinders. The next model had a 525-cubic-inch engine with six cylinders.

They became a West Virginia company when they got into financial trouble in Ohio and moved to Martinsburg, West Virginia, in 1911. They stayed there until the company's demise in 1922.

The company's most famous model was the *Norwalk Underslung Six.* This car was advertised as "A Car of Absolute Exclusiveness." It had forty-inch wheels, a silk mohair top, and a full electrical system powering the starter, lighting, horn, cigarette lighter, and more. The $3,000 price for the 1912 two-passenger roadster or the six-passenger touring model was enough to keep the company going until 1922.

The Norwalk of West Virginia

Chapter 49:
The Jeffery, the Mitchell, and the Wisconsin Car Companies

THE **Mitchell Motor Company** of Racine, Wisconsin, started off as a wagon-building company in 1834 and in 1903 started building automobiles. The combined wagon and car company was called the **Mitchell-Lewis Company, Ltd.** They had a building at 815 8th Street in Racine.

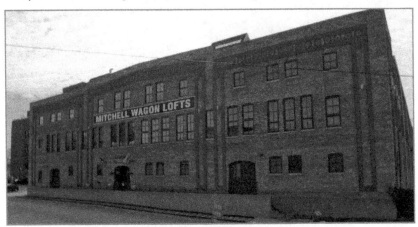

The Mitchell-Lewis Building

Mitchell-Lewis Motor Company,
Racine, Wisconsin

Notable *Mitchells* included the 1920 *Model E-40*, which had a sloping radiator that caused the auto to be called the "Drunken Mitchell." *Mitchells* were known as large, sleek luxury vehicles and were available in four-, six-, and eight-cylinder engines.

In 1923, the company went under, and the remains were purchased by the **Nash Motor Company**.

A 1903 Mitchell automobile, the first year for the marque

1912 Mitchell

1903 Rambler

The **Thomas B. Jeffery Company** of Kenosha, Wisconsin, is one of the oldest of the auto companies, starting in 1902 and ending as part of the **American Motors Company** from 1954 to 1988.

The *Jeffery* was named after auto pioneer Thomas B. Jeffery by his son Charles Jeffrey, replacing the famous *Rambler* in 1914. The company manufactured 40-horsepower, four-cylinder cars and also a six-cylinder car called the *Chesterfield Six*. Production for 1914 was 10,417 *Jeffrey* cars and 3,096 units of a truck called the *Jeffery Quad*.

Rambler logo and 1902 Rambler

In 1915, the firm concentrated on trucks, producing over 7,600 of them, and only about three thousand of the popular *Jeffrey*. In 1916, a two-passenger all-weather model was offered. However, in May 1916, Charles Jeffrey took a trip on the ill-fated *Lusitania*, and though he was one of the survivors when it sank, he spent four hours in the icy water waiting to be rescued. Right then, he decided that life was too short and once rescued, he retired at the age of forty. The 1917 *Jeffery*s were the last ones. Charles Jeffrey sold the company to Charles Nash of the **Nash Motor Company**.

Pierce-Racine

Pierce-Racine radiator badge

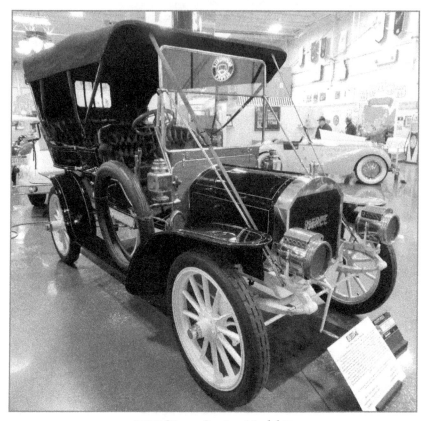

1907 Pierce-Racine Model D

Andrew J. Pierce of Racine, Wisconsin, started a company in 1894 called the **Pierce Engine Company**, with marine engines as a specialty. Pierce built his first two-cylinder auto in 1895, his second in 1899, and a third in 1901. He worked for the **Mitchell Motor Company** for a short while and then formed his own company, the **Pierce Motor Company**.

In 1904, Pierce's first car, the *Pierce-Racine,* was a one-cylinder, 8-horsepower runabout model. It had a planetary transmission, chain drive, and a steering wheel when most cars still used a tiller. In 1905, he doubled down with two-cylinder, 16-horsepower vehicles. Models available were runabouts and tonneaus. In 1906, the company released a four-cylinder touring car. The company's best year was 1910, when over one thousand units were sold.

Some of the car's investors were men from the **J.J. Case Threshing Machine Company** and in 1911 engineered a takeover of the company. The last *Pierce-Racine*s were released in 1911, and thereafter, the car was called a *Case*. Over two thousand *Pierce-Racine*s were manufactured.

The **J.J. Case Threshing Machine Company** of Racine, Wisconsin, manufactured a car called the *Case* from 1910 to 1927. They also made steam-traction engines, threshing machines, and tractors. The cars were sold through the same agricultural dealers who sold farming equipment in the New York area. Only four-cylinder autos were offered until 1916, when a car with a six-cylinder Continental engine was made available.

When the company ceased auto production in 1927, they continued to make agricultural farm equipment.

The 1918 Case

Louis Kissel and his four sons formed the **Kissel Motor Car Company** in Hartford, Wisconsin, in 1906. They had, prior to forming the car company, been known for owning a hardware store and manufacturing agricultural equipment.

Their first auto, named the *Kissel Kar*, was a 35-horsepower, four-cylinder driven auto. (It was at first called the *Badger*.) Most parts, unusual for the time, were made by the Kissel's company.

The Kissell Motor Car Company factory, outside and inside

Notably, the car got rid of the crank and got electric starters in 1913. In 1917 and 1918, the company built a V-12 engine. The tone of the company changed to more of a sporty image in 1918, as the *Kissel Kar Silver Special Speedster* would demonstrate. (The "silver" part of the name was after designer C. T. Silver.)

In January 1918, the firm's most famous car, the *Kissel Gold Bug,* was released. Its most predominant color was gold-hued yellow, hence the *Gold Bug* name. Models such as the *Coach-Sedan* and the *Tourster* were added, and in 1924, Lockheed hydraulic brakes were an option. Another option was a straight-8 Lycoming engine.

The 1929 *White Eagle* had six- or eight-cylinder engines as an option and standard hydraulic brakes, but the $1,595 to $3,885 prices were not competitive, and they produced only 1,531 autos. The *Ruxton* was assembled in the **Kissel** factories in 1930, and the company went into receivership in 1931 to avoid a takeover by Archie Andrews. They reorganized in 1933 to manufacture Lever engines.

1923 Kissel Gold Bug Speedster

1925 Kissel Gold Bug

Chapter 50:

The Wyoming Cars of Elmer Lovejoy

In the winter of 1897–1898, Elmer Lovejoy built Wyoming's first automobile in his Laramie bicycle shop. He attached a two-cylinder, two-cycle marine engine to a carriage, constructing what the local newspaper called a horseless carriage. After a successful test drive on May 7, 1898, Lovejoy talked about marketing the car to the masses. It ran at two speeds: five and ten miles per hour. His car lasted two years and then stopped running.

Lovejoy went on to become a *Franklin* auto dealer and invented a steering mechanism that is still used today. It is a steering spindle which permits each front wheel to pivot. Because he didn't have the $350 needed to have a patent registered, he sold the rights to the device to the **Locomobile Company of America**. He received $800 and a new *Locomobile Steamer* car for it. In 1917, he invented the automatic door opener and manufactured them for several years. In the 1930s, he had an auto repair shop. The Lovejoy Building in Laramie is named after him.

Lovejoy's Garage

CHAPTER 51:

THE CAR COMPANY OF WASHINGTON, D.C.

THE **JENKINS AUTOMOBILE COMPANY** AT 1325 F Street in Washington, D.C., first built a steam car with 2½ horsepower, headed by C. Francis Jenkins in 1900. Jenkins went on to make a freight truck and a large, passenger-carrying coach, both powered by steam.

In 1901, he built an electric vehicle he called the "littlest auto ever made." It was a Victoria-style automobile with a twenty-four-inch wheelbase, built for a Cuban midget named Chiquita. Chiquita drove the car in the Pan-American Exposition of 1901. The **Jenkins Automobile Company** appears to have gone dormant in early 1902.

Chapter 52:

Car Companies of Multiple States

EVERAL CAR COMPANIES AND MODELS were situated in two (or more) different states. Some of these were the **Sellers Motor Company**, which was in both Indiana and Kansas; the **Riker Electric Vehicle Company**, which was in both New York and New Jersey; the *Baldwin Steam*, which was in both Rhode Island and Pennsylvania; the **Abbott Motor Company**, which was located in Detroit and then Cleveland; the *Jonz*, which was in both Indiana and Nebraska; the *Crouch*, which was in both Pennsylvania and Maryland; and the *Bendix*, which was in both Chicago, Illinois, and Logansport, Indiana.

The 1912 Empire Runabout, the 1912 Empire Touring Car, and the 1913 Empire

In 1909, Arthur Newby and Robert Hassler of the **National Motor and Electric Company** (which produced the *National* in Indianapolis), **Prest-O-Lite Starters** founder Carl Fisher, and James Allison of aircraft fame combined forces to build the *Little Aristocrat,* a four-cylinder, 20-horsepower roadster, selling for $850. Production began in 1910, and 289 cars were built. The owners were also the founders of the Indianapolis oval racetrack that became the host of the Indianapolis 500, the famed Memorial Day car race. The first car to run the track was the **Empire Motor Car Company**'s *Little Aristocrat.*

In 1911, the company hired Harry Stutz, who changed the cars from chain drive to shaft drive before leaving to start his own company, but the car company came second in the priorities of the racetrack owners, and in late 1911, they sold the company to a group who reorganized the company as the **Empire Automobile Company.**

As the **Empire Automobile Company**, the reborn firm needed a place to build their auto since their original factory was being used to build **Prest-O-Lite** starters. In 1912, they found the factory of the **Greenville Metal Products Company** in Greenville, Pennsylvania, and built both the *Empire* and a similar car called the *Fay* there (named after Frank Fay, the owner of the factory).

In 1915, they acquired the factory of the defunct **Federal Motor Company** in Indianapolis and moved operations there. The car, no longer called the *Little Aristocrat,* was now available in a four- or six-cylinder touring car. It was called the *Empire,* followed by the name, including the *Empire 20* and the *Empire Model 25.*

The last year the **Empire Automobile Company** released a new car was 1919. Before going under, the company produced a little over 4,800 cars, though few remain.

———————————

You knew the automobile had definitely arrived when you could buy one in the Sears catalog! The year that happened was 1908. The *Sears* was a high-wheeler model with a two-cylinder, 10-horsepower, gasoline engine. It had a friction transmission and a double-chain drive. The first ones were built in the **Hercules Buggy Factory** in Evansville, Indiana, but it moved production to its new factory at Harrison and Loomis Street in Chicago as soon as it was ready. The first car sold by the **Sears Motor Car Works** cost $395 and was a solid-tired runabout model.

In 1910, five auto models and a light delivery vehicle made their way to the Sears catalog. Successive models up the alphabet added more to the car:

- *Model G* was the basic, no-frills Sears auto
- *Model H* had a top and fenders
- *Model J* had a top, fenders, and running boards
- *Model K* included everything plus cushion tires
- *Model L* had everything plus pneumatic tires
- *Model M*, in 1911, had a "cozy coupe top"

The cars were discontinued in 1912 after over 3,500 were sold because the company was losing money on each one. The **Sears Motor Car Works** machinery was sold to the **Lincoln Motor Car Company**.

1910 Sears

In Chicago from 1910 to 1912, the **Kenmore Manufacturing Company** produced mostly two-cylinder cars in surrey, runabout, and roadster models. Sears bought out the company in 1912 and used the Kenmore name for appliances from then on.

When Henry J. Kaiser couldn't sell his *Henry J* auto fast enough, he hit upon the idea of using the resources of the Sears network. Rebranding the car (literally), they put *Allstate* badges on the car and in 1951 and 1952 sold it through the Sears catalog.

1952 Allstate auto

The New York distributor of the **Rauch & Lang Company**, Clyde Kenworthy, felt that electrics were going out of vogue and wanted to find a gasoline car to sell. When he couldn't find a good one, he joined with Albert Barley, who had just designed the *Halladay*, and Karl Marting, who designed the *Wasp*.

Together, they all designed the large and limo-like *Roamer* in 1916, named after a famous racehorse of the day. They formed the **Barley Motor Car Company** in Streator, Illinois, to assemble the *Roamer*, a six-cylinder, 23-horsepower touring car. When Kenworthy and Marting dropped out, Barley divested his **Halladay Motor Car Company** interests in Streator, reorganized the company, and moved to an old factory of the **Michigan Buggy Company** in Kalamazoo, Michigan, in 1918.

1921 Roamer

From 1918 on, Duesenberg engines were used. The 1919 *Roamer* was available as a touring car as well as cabriolet, town car, touring sedan, standard sedan, and roadster models, all selling in the $2,500 range. In 1921, a *Roamer* set a world stock chassis speed record of 105.08 miles per hour. Mary Pickford and Buster Keaton both owned *Roamers*.

From 1922 to 1924, the company manufactured a less expensive version of the *Roamer*, the *Barley*. The *Barley* had a six-cylinder engine and was available in sedan or torpedo models. It sold for $1,400, more with options.

For 1924 and 1925, *Barley* turned Kalamazoo into the taxicab capital (Kalamazoo also had the **Checker Cab Company**) of America and manufactured a taxi called the *Pennant*.

After discontinuing its subsidiaries, the last *Roamers* rolled off the assembly line in 1929, having produced over twelve thousand vehicles over the life of the marque.

The **Rauch & Lang Company** started as the **Rauch & Lang Carriage Company** in 1865 in Cleveland, Ohio, and were known for making some of the most luxurious carriages of the day. In 1903, they bought the **Buffalo Electric Car Company** and started making *Rauch & Lang Electric Cars*. By 1908, they were manufacturing over five hundred vehicles a year.

In 1911, the **Rauch & Lang Company** was sued by the **Baker Motor Vehicle Company** for infringing on their rear springs patent. Faced with declining sales, in 1915, the two companies settled by merging and becoming the **Baker R & L Company**.

The Owen brothers, Raymond M. and Ralph R., had a long and varied history in the automotive world in many different states. They first started working together as the **R.M. Owen**

Carpet Cleaning and Rug Manufacturing Service of Cleveland. In 1899, they attached an 8-horsepower gasoline engine (of their own invention) to their delivery wagon. When the vehicle proved a hit with the neighbors, the brothers formed a company to build the delivery wagons. They advertised that a seat could be added to make it a passenger vehicle. At first, they called themselves the **Phoenix Motor Vehicle and Engine Company** and then changed it to the **Owen Motor Carriage Company**. The vehicles were therefore known as both *Phoenix*es and *Owen*s.

Owen factory

In 1901, the brothers were first impressed by Ransom E. Olds and *Oldsmobile*. Raymond moved to New York City to set up an *Oldsmobile* distributorship, and Ralph went to Lansing, Michigan, and became factory manager of the **Olds** factory. When Ransom Olds made his exit from the **Olds Motor Company** in 1904 to form the **REO Motor Company**, Raymond also changed his New York distributorship to the **REO Motor Company**.

In 1910, Ralph got some auto friends from the **Olds Motor Company** together to form a new auto company in Detroit. They incorporated as the **Owen Motor Car Company** with $500,000 in capital. In 1911, they released their first auto, a four-cylinder, 50-horsepower, seven-seater touring car, which sold for $4,000. The car used a central gear change, one of the first autos to have this feature. When sales for the vehicle were slow, Ralph appealed to his brother Raymond, the **REO Motor Company** dealer. Ransom Olds bought out the car company in 1911 and assembled the last thirty-five autos before discontinuing the brand. The **Krit Motor Company** took over the **Owen Motor Company** factory.

The next appearance of the Owen brothers on the auto-manufacturing landscape was when they appeared at 142nd Street and Fifth Avenue in New York City in 1915. They were back as **R.M. Owen & Company** with their most renowned vehicle, the *Owen Magnetic*. This was an electric vehicle, using the electric transmission demonstrated in the *Entz Six* in the 1914 New York Automobile Show. The Owens had the *Owen Magnetic* ready for the 1915 New York Auto Show.

The *Owen Magnetic* was a hybrid vehicle that used a six-cylinder gasoline engine to charge a generator that created an electric magnetic field used to power the car's electric, five-speed, automatic transmission. A popular luxury car, famous owners included Enrico Caruso and John McCormack.

The **Baker Electric Company** and the **Rauch & Lang Company**, two electric car companies, joined in a consortium with **R.M. Owen & Company**, prompting a move from New York, where 250 cars were manufactured, to Cleveland, Ohio. Production resumed as the *Owen Magnetic*'s engines and chassis were built at the **Baker** plant, and the **Rauch & Lang** plant was used for the *Owen*'s coachwork. In 1916, a sporty touring model was added, priced at $6,000. When the **Baker** factory started war production in 1919, Raymond Owen moved production of the car to the former *Matheson* factory in Forty-Fort, near Wilkes-Barre, Pennsylvania, from 1919 to 1922. In 1922, when the *Owen Magnetic* was no longer being produced in the United States, Raymond Owen manufactured it in

England under the name *Crown Magnetic.*

The *Matheson* started in Grand Rapids, Michigan, in 1906, moved to Holyoke, Massachusetts, and then settled in Wilkes-Barre, Pennsylvania, until its demise in 1912. In 1903, brothers Frank and Charles Matheson both worked for the Fred Macey Company, a mail-order furniture business, and spent their time experimenting with auto engines. (Grand Rapids was the "furniture capital" of the United States.) The brothers heard that the **Holyoke Motor Works** was for sale in Holyoke, Massachusetts, and they purchased the company in 1904. Along with the motor company, they inherited engineer Charles Greuter. Greuter had developed an overhead valve engine that the brothers wanted to use in their auto. The new company was named the **Matheson Motor Car Company, Ltd.**

Tom Cooper in a Matheson race car

At first, the cars were assembled in Grand Rapids, and the engine and transmission work were done in Holyoke, Massachusetts, but in 1904, the whole operation was moved to Holyoke. In the company's first year, they manufactured and sold sixty autos—which at $5,000 per car was a good start for the company.

1908 Matheson Landaulet

In March 1906, the company was invited by the town of Wilkes-Barre, Pennsylvania, to relocate there. By the time of the *Matheson Super Six,* the *Matheson* was a formidable race car, driven by such famous drivers as Barney Oldfield and Ralph DePalma, who took the car to victory in several races from 1906 to 1912. American Automobile Association president Charles Whipple loved his *Whipple*, and so did William Randolph Hearst, who purchased two of the autos. Bicycle racer and early investor of Henry Ford Tom Cooper met his end in one.

The company went into its first receivership in 1910 but made their way out. In 1911, their most famed auto, the *Silent Six,* was released and then again in 1912, which is when the company went out of business.

The assets were sold in April 1913, and Frank Matheson stayed until 1914, helping to convert the factory to wartime production. After World War I, he joined Raymond Owen in producing the *Owen-Magnetic* automobile until the company's failure in 1921. Matheson then bought his old *Matheson* plant back to use as a distributorship for *Dodge, GMC,* and *Oakland* (later *Pontiac*) models until it burned down in 1926. Charles Matheson became the vice-president of sales at the **Dodge Brothers Motor Company** and then moved on to be vice-president of sales at the **Oakland Motor Company** when they were making the transition to becoming **Pontiac**. He later launched the sales campaign for the *DeSoto,* and he was general sales manager at the **Graham Motor Company** when he died in 1940.

The *Tincher* was built in Chicago and made its debut at the 1903 Chicago Auto Show. It made a sensational debut chiefly due to its air pump that operated the first power brakes in an automobile. The air pump could also be used to pump air into the tires and to toot the car's horn!

The 1907 Tincher

The *Tincher* was the brainchild of Thomas Tincher and offered 90-horsepower, 126-inch wheelbase, luxury autos with four-speed transmission and double-chain drive. Although in 1903 it was a four-cylinder auto, by 1908, a six-cylinder model was manufactured.

In 1907, the company moved to South Bend, Indiana, and incorporated as the **Tincher Motor Car Company**, with the Studebaker brothers as the majority stockholders. With roadsters, touring models, and race cars, the *Tinchers* sold from $5,000 to $12,000 per car, but the cars had too low of a production to be successful financially. In 1909, Thomas Tincher declared bankruptcy and went on to work for the **Economy Motor Buggy Company** in Joliet, Illinois.

The *Federal* was designed in Chicago by John Waters in 1907 and manufactured at first by the **Federal Motor Car Company** and then by the reorganized **Federal Automobile Company**. They were reorganized again in 1908 and moved to Rockford, Illinois, as the **Rockford Automobile & Engine Company**, and the car was briefly called the *Rockford*. A final reorganization occurred in 1909 when the whole company was purchased by the **Industrial Automobile Company** and moved to Elkhart, Indiana, where very few cars were manufactured before the company went under.

The auto they produced was a high-wheeler, seating two passengers, with a two-cylinder, two-stroke, air-cooled engine selling for $600. Features included fenders and a folding top, and their motto was, "An Ideal Car for Business and Professional Men." Solid tires were standard equipment, and pneumatic tires could be bought for an additional $100.

Henry Nyberg built his first auto, a two-cylinder runabout, in 1898 right after he graduated from technical school in Malmo, Sweden, and immigrated to the United States. At first, he worked at **Jeffries Auto Works** in Kenosha, Wisconsin. He then started a partnership called the **Nyberg-Waller Automobile Company** in Chicago.

In 1904, the growing auto company moved out of Nyberg's machine shop and into a larger facility at 30th Street and Michigan Avenue in Chicago. Car repair took up most of their time, and not many cars were manufactured. In 1907, Nyberg and Waller parted ways, and Nyberg joined with H. E. Jennings to reorganize as the **Nyberg Automobile Company**. Their stated purpose was to "reconstruct and sell secondhand cars."

In 1908, the company announced it would soon manufacture their own car, but it never happened for Nyberg until 1911, when he moved to Anderson, Indiana, after purchasing the **Rider-Lewis Motor Company**. He set up production in their old factory for his handmade car, the *Nyberg*. It had a four-cylinder, 40-horsepower engine and was available in touring and roadster models.

In 1912, the company released a six-cylinder model, the *Nyberg 6*. When the plant had seventy employees, Nyberg opened another plant in Chattanooga, Tennessee. The company made not only cars but also trucks and fire engines, but undercapitalization doomed his ventures in 1913. The company was forced into receivership; Nyberg blamed the spring rains in Anderson, Indiana.

A. C. Barley of Streator, Illinois (who manufactured the *Hall-aday* auto) bought out the business and, in 1915, brought out the *Madison,* designed by Henry Nyberg. After Barley, Nyberg was later associated with Cecil Gibson of the **Empire Company of Indianapolis** and then retired in Sweden.

The factory in Anderson, Indiana, where the Nyberg, Rider-Lewis Cyclecar, and Madison autos were built (photo from the Library of Congress)

Henry Nyberg in a 1912 Nyberg roadster

*1912 Dixie Special manufactured by **Nyberg Automobile Works**.*
The factory was located at George and Buckley Streets
in Chattanooga, Tennessee.

Brewster & Company of Long Island, New York, started in the carriage business in 1810. In 1915, they released a vehicle available in several models with a four-cylinder, 26–53 horsepower Knight engine, a high-priced auto called the *Brewster*. It came in a variety of styles, including a town landaulet that cost $8,450, a glass-quarter brougham for $8,300, and a falling-front landaulet for $8,500. They continued making cars until 1925, when they started making bodies for **American Rolls Royce** in Springfield, Massachusetts. The company was absorbed into **American Rolls Royce** in 1926.

The *Brewster* made a comeback in 1934, as **American Rolls Royce** changed their name to the **Springfield Maufacturing Corporation**, but this time, the Great Depression did them in by 1935.

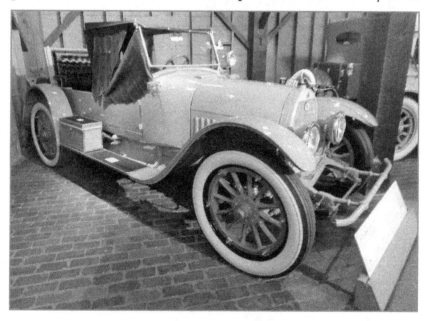

1915 Brewster

The **Abbott Motor Company** started in Detroit in 1909 with a four-cylinder, 30-horsepower touring car. A popular car, it won events like endurance and hill climbing. Producing over ten thousand cars in its career, the car was at first named the *Abbott-Detroit*.

Abbott had a revolving management starting with founder Charles Abbott, who resigned less than a year after forming the company. Since sales were good, in 1916, the company decided to take a ten-year lease on a factory in Cleveland. Cars manufactured in Cleveland were just called *Abbotts* without the *Detroit* appellation. Unfortunately, sales were not enough to pay the rent, and the company became insolvent in 1917.

Abbott factory in Cleveland

Lou Halladay designed a four-cylinder, 40-horsepower vehicle in Anderson, Indiana. Buying out the **Erie Motor Carriage and Manufacturing Company,** he moved the whole operation to Streator, Illinois. Once in Streator, he rounded up capital ($30,000) and organized the **Streator Automobile and Manufacturing Company,** soon shortened to the **Streator Motor Car Company.**

The first auto came out in 1905, the *Halladay,* and was at first available only in a touring model, but limo and runabout models were added the following year. They used the motto "Every Day a *Halladay*" and were manufacturing around nine hundred cars a year.

In 1911, the company had financial problems and went into receivership. The company ended up in the hands of Albert Barley, who reopened the factory and again began to manufacture the *Halladay.*

Things went well until 1916, when Barley became involved with the *Roamer,* sold his interest in the company, and moved to Kalamazoo. The financial group he sold it to relocated the operations to Ohio, using manufacturing facilities in Warren, Attica, and Newark to continue making the *Halladay* brand. They renamed the company the **Halladay Motor Car Company** and later the **Halladay Motors Corporation** and continued manufacturing until 1922, when they went into receivership.

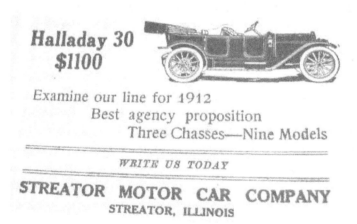

Halladay 30
$1100

Examine our line for 1912
Best agency proposition
Three Chasses—Nine Models

WRITE US TODAY

STREATOR MOTOR CAR COMPANY
STREATOR, ILLINOIS

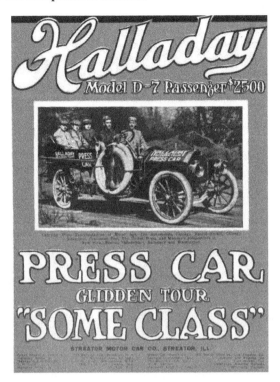

GLOSSARY

Accumulator: British for car battery.

Antique Era: A designation for any auto before the 1920s; this would include Brass Era and Veteran's Era autos.

Bonnet: British for a car's hood.

Brass Era: The period between 1890–1919 (1896–1915 according to some), when autos were in their infancy. Named after the number of brass fittings used in early autos.

Brougham: A closed, four-wheel buggy or carriage with an open driver's area; any vehicle with an open driver's area; an electric car that resembles a coupe.

Cabriolet: A convertible car with windows, usually a two-door model.

Coupe: A closed, four-wheel buggy or carriage with an open driver's area; a closed auto with two doors and a fixed roof.

Cyclecar: A vehicle in between a car and a motorcycle. Usually a one- or two-cylinder vehicle, sometimes with only three wheels. The small cars were narrow and sold from $250 to $500; they were popular in the 1910s until the Ford *Model T* eclipsed them in popularity.

Cylinder: A cylindrical chamber in which the pressure of a gas or liquid moves a sliding piston. Piston rings clamp the cylinder inside a piston in a car engine. Usually arranged in rows, the number of cylinders determine the size and power of an engine.

High-wheeler: A type of auto style between 1890–1910 in the Brass Era of autos. It was descended from and looked like buggies and horse-driven carriages. With large, slender wheels, they could provide clearance on the substandard roads of the day.

Horseless Carriage: Term for autos built before 1910.

Horsepower (hp): First used by James Watt in the late 1700s to compare the output of work horses to steam engines, it is officially "a foot-pound-second unit of power equivalent to 550 foot-pounds per second, or 745.7 watts."

Landau: Originally a carriage term, for modern vehicles, it refers to a "simulated" convertible; a car that looks like it has a removable top but really doesn't.

Landaulet: Originally a carriage style, now it is like a limousine model but with a convertible top instead of a permanent one over the passenger section (so it can be removed). This type is used in parades for public figures to ride in and be seen.

Marque: Brand or make of an auto.

Motor car: An early term for an automobile.

Phaeton: A four-door, open touring car.

Post-Vintage Era: A designation for cars built in the 1930s.

Prototype: The first, preliminary construction of an auto. It is the model that is copied for other cars that are subsequently built.

Raceabout: A one- or two-seater auto with race car lines and styling.

Roadster: A sports car style (usually roofless) with one seat for two in the front and a small luggage compartment in the back.

Rumble Seat: Also called the "mother-in-law seat," this was an open, fold-up car seat in the rear trunk area of the car.

Runabout: A small, light auto with an open top capable of carrying two passengers.

Running Board: A metal strip that ran between the fenders and below the doors of early autos and trucks, often used as a step or to jump onto while the car was running so one could shout, "Follow that car!"

Sideboard: Ornamental receptacle in a car usually facing the central compartment of the car.

Surrey: A four-wheeled carriage with two seats able to seat four people. A similar design was used for early motor cars.

Steamer: A car with a steam-powered engine.

Stanhope: A type of open horse carriage adapted to autos, this was a light, open, one-seater car.

Stroke: The piston completes separate strokes while turning the crankshaft. A stroke refers to the full travel of the piston along the cylinder in either direction.

Tonneau: On Brass Era cars, the rear seat compartment, or a removable cover for the seating area of an early auto.

Torpedo: A type of body style with hood line raised to be level with the car's waistline, resulting in a straight beltline (the demarcation between a vehicle's body panels and the side windows) on an automobile from front to back, now replaced by the hardback.

Tourabout: Like a touring car but using two separate seat sections.

Touring Car or **"Tourer":** An open car with a folding top, usually without windows, seating five to seven, including the driver.

Town Cabriolet: A town car in which the covered rear section can be converted to an open car.

Town Car: A chauffeur-driven auto where the passengers are fully enclosed and the driver is fully exposed.

Toy Tonneau: A form of rear-seat arrangement for two passengers only. The difference from a "regular" tonneau is that the backs of the seats are much lower than in the standard type.

Veteran Era: Automobiles assembled before the year 1890.

Vintage Era: A designation for cars manufactured between 1919–1930.

Wheelbase (wb): The distance from the center of the front-wheel rod to the center of the rear-wheel axle, usually given in inches in the United States, or, more simply, "the distance between the front and back axles of a motor vehicle," a basic determinant in how large a car is.

Bibliography

Anderson, Rudolph E. (1950). *The Story of the American Automobile*. Washington, DC: Public Affairs Press.

Bausch, David K. (1996). *Official Price Guide to Automobilia*. New York, NY: House of Collectibles.

Beck, Lee & Malko, Josh B. (1996). *Auburn & Cord*. Osceola, WI: Motorbooks International Publishers & Wholesalers.

Burness, Tad (2001). *The Ultimate Auto Album*. Iola, WS: Krause Publications.

Butler, Don (1992). *Auburn, Cord, Duesenberg*. Jefferson, NC: McFarland & Company, Inc., Publishers.

Crabb, Richard (1969). *Birth of a Giant: The Men and Incidents That Gave America the Motorcar*. Philadelphia, PA: Chilton Book Company.

Clymer, Floyd (1953). *Those Wonderful Old Automobiles*. New York, NY: Bonanza Books.

Clymer, Floyd (1971). *Treasury of Early American Automobiles*. New York, NY: Bonanza Books.

Einstein, Arthur W. (2010). *"Ask the Man Who Owns One": An Illustrated History of Packard Advertising*. Jefferson, NC: McFarland & Company, Inc., Publishers.

Flinn, Gary (2017). *The Hidden History of Flint*. Charleston, SC: History Press.

Foster, Patrick K. (2013). *American Motors Corporation*. Minneapolis, MN: MBI Publishing Company.

Georgano, G. N. (ed.) (1968, 1972). *The Complete Encyclopedia of Motorcars, 1885 to the Present*. New York, NY: E.P. Dutton and Company, Inc.

Glasscock, C. B. (1937). *Motor History of America*. Los Angeles, CA: Floyd Clymer Publishing.

Goldstone, Lawrence (2016). *Drive!: Henry Ford, George Selden, and the Race to Invent the Auto Age*. New York, NY: Random House.

Hamper, Ben (1992). *Rivethead: Tales from the Assembly Line*. New York, NY: Warner Books.

Heaton, Dan (2012). *Forgotten Aviator: The Byron Q. Jones Story*. Boston, MA: Branden Books.

Henshaw, Peter (2010). *The Encyclopedia of Cars: An Illustrated Guide to Classic Motorcars With 600 Photographs*. Dayton, OH: Lorenz Educational Press.

Horvath, Dennis E. & Terri (1997). *Cruise IN: A Guide to Indiana's Automotive Past and Present*. Indianapolis, IN: Publishing Resources.

Hudson Triangle. (2010). Ann Arbor, MI: University of Michigan Library.

Hyde, Charles K. (2009). *Storied Independent Automakers*. Detroit, MI: Wayne State University Press.

Hyde, Charles K. (2005). *The Dodge Brothers: The Men, the Motor Cars, and the Legacy*. Detroit, MI: Wayne State University Press.

Kimes, Beverly Rae & Clark, Henry Austin Jr. (1985). *Standard Catalog of American Cars 1805–1942*. Iola, WS: Krause Publications.

Kimes, Beverly Rae & Clark, Henry Austin Jr., et al. (1985). *Standard Catalog of American Cars 1945–1975*. Iola, WS: Krause Publications.

Kimes, Beverly Rae (1976). *Great Cars & Grand Marques*. New York, NY: Bonanza Books.

Langworth, Richard M. (1987). *The Complete History of Ford Motor Company*. New York, NY: Beekman House.

Lehto, Steve (2016). *Preston Tucker and His Battle to Build the Car of Tomorrow*. Chicago, IL: Chicago Review Press.

Long, J. C. (1945). *Roy D. Chapin: The Man Behind the Hudson Motor Car Company*. Detroit, MI: Wayne State University Press.

Ludvigsen, Karl & Wise, and David Burgess (1982). *The Encyclopedia of the American Automobile. Exeter, N.H.: Exeter Book Publishing Company*.

Madden, W. C. (2003). *Haynes-Apperson and America's First Practical Automobile*. Jefferson, NC: McFarland & Company, Inc., Publishers.

Mandel, Leon (1982). *American Cars*. New York, NY: Stewart, Tabori, & Chang, Publishers.

May, George S. (1975). *A Most Unique Machine: The Michigan Origins of the American Automobile Industry*. Grand Rapids, MI: William B. Eerdmans Publishing Company.

May, George S. (ed.) (1990). *The Automobile Industry, 1896–1920*. New York, NY: Facts on File.

May, George S. (ed.) (1990). *The Automobile Industry, 1920–1980*. New York, NY: Facts on File.

McConnell, Curt (1995). *Great Cars of the Great Plains*. Lincoln, NE: University of Nebraska Press.

Miller, G. Wayne (2015). *Car Crazy: The Battle for Supremacy Between Ford and Olds and the Dawn of the Automobile Age*. New York, NY: PublicAffairs Books.

Nelson, Dean (2016). *Byron J. Carter*. Minneapolis, MN: Cyclorium Press.

Olsen, Byron & Cabadas, Joseph (2002). *The American Auto Factory*. St. Paul, MN: Motorbooks.

Powell, Sinclair (1999). *The Franklin Automobile Company*. Warrendale, PA: Society of Automotive Engineers International.

Stanley, Richard A. (2012). *Custom Built by McFarlan: A History of the Carriage and Automobile Manufacturer 1856–1928*. Jefferson, NC: McFarland & Company, Inc., Publishers.

Szudarek, Robert G. (1996). *How Detroit Became the Automotive Capital*. Detroit, MI: Typocraft Company.

Coppolla, Francis Ford. (1988) Tucker: The Man and His Dream (DVD). San Francisco, CA: Zoetrope

Ward, James A. (2003). *Three Men in a Hupp: Around the World by Automobile, 1910–1912*. Palo Alto, CA: Stanford University Press.

Wrenick, Frank E. & Elaine V. (2014). *Automobile Manufacturers of Cleveland and Ohio, 1864–1942*. Jefferson, NC: McFarland & Company, Inc., Publishers.

ABOUT THE AUTHOR

S A COLLEGE STUDENT, ALAN Naldrett started one of the nation's first used record stores on the campus of Michigan State University, where he received his BA degree. Living in California as an insurance and stock broker, he relocated back to Michigan, where he acquired master's degrees in library and information science and archival science from Wayne State University. He was first a medical librarian and then an academic librarian for a private Michigan college for more than ten years.

Upon retirement, he continued writing books, a practice he started while a reference librarian. He has now written over fifteen books including *Lost Towns of Eastern Michigan, Lost Car Companies of Detroit,* and *Michigan's C. Harold Wills—The Genius Behind the Model T and the Wills Sainte Claire Automobile.* Another action Alan has taken up is archival organization. He has arranged historical files for colleges, townships, schools, churches, and museums. Alan is the former vice-chairman of the Macomb County Historical Commission and, presently, vice-president of the Chesterfield Township Library Board of Trustees. A member of the Michigan Historical Society, he has written research articles for *Michigan History* and *Michigan Chronicle* magazines. A life member of the Chesterfield and New Baltimore Historical Societies and the Wills Sainte Claire Museum, he has conducted many programs and presentations for history lovers all over Michigan. He was presented the Award of Merit by the New Baltimore Historical Society and the Spirit of Wills Award by the Wills Sainte Claire Museum. He is also a member of the F Street Blues, a blues band.

Alan can be reached at alannaldrett@gmail.com.

Lots of Studebakers-Courtesy of the Studebaker Museum, South Bend, Indiana